1252

Freshwater Biology

Studies in the Biological Sciences
General Editor: *Professor V. Moses*

The Biology of Insects: *C. P. Friedlander*
Microbial Plant Pathology: *P. J. Whitney*

Freshwater Biology

L.G. Willoughby

Pica Press   New York

Published in the United States of America in 1977 by
PICA PRESS
Distributed by Universe Books
381 Park Avenue South, New York, N.Y. 10016

© L. G. Willoughby 1976

All rights reserved. No part of this publication
may be reproduced, stored in a retrieval system, or
transmitted, in any form or by any means, electronic,
mechanical, photocopying, recording, or otherwise
without the prior permission of the publishers.

Library of Congress Catalog Card Number: 76-20405
ISBN 0-87663-721-7

Printed in Great Britain

Contents

	Preface	7
1	**The physical and chemical background**	9
	Temperature	9
	Dissolved oxygen	12
	Calcium	14
	Nitrates and phosphates	15
2	**Photosynthetic plants**	17
	Algae	19
	Phytoplankton	21
	Comparative phytoplankton studies	26
	Benthic algae	28
	Algal blooms	31
	Nitrogen fixation by blue-green algae	32
	Algal toxins	35
	Algal viruses	36
	Aquatic vascular plants	38
	Aquatic vascular plants as nuisance organisms	42
3	**Invertebrate animals**	46
	Protozoa	46
	Crustacea	49
	Insecta	52
4	**Fish biology**	61
	The pike, *Esox lucius*	61
	Salmonid fish	74
	Salmon	74
	The Pacific salmon	75
	Further aspects of fish feeding	81
	Fish introductions	82
	Fish ponds	84
	The introduction of food organisms	85
	Heated effluents	88
	Serology	88
	Fish diseases	89

5	**Streams and rivers**	93
	Gammarus	98
	Sampling the invertebrate fauna	100
	River quality	101
6	**Decomposition cycles and nutrient re-circulation**	104
	The role of the fungi	104
	The role of the bacteria	118
	Soluble phosphorus circulation	129
7	**Lakes, reservoirs and water supplies**	132
	Water treatment	133
8	**Sewage and industrial waste disposal**	137
9	**Pollution phenomena**	142
	Pesticides	146
	Fish and river pollution	147
10	**The sedimentary record**	149
	Sedimentation interpretation	149
	Chemical evidence of biological evolution	150
	Plant hydrocarbons	152
	Plant pigments	152
	Animal remains	153
	Very recent deposits	154
	Summary	156
11	**The way ahead**	158
	Mosquito control	158
	Fish farming	159
	Reservoir biology	159
	Index	163

Preface

In this book I have attempted to blend a general treatment of the subject together with discussion of some of the more recent researches, and their objectives. The amount of this material which might have been drawn on is so enormous that some selectivity has been inevitable. However, my hope is that a reasonably balanced picture will emerge. I am grateful to Dr Winifred E. Frost and Miss Charlotte Kipling for reading Chapter 4 and making suggestions for its improvement, also to Professor G.W. Minshall for guidance in invertebrate zoology.

Permission to reproduce illustrations, many in a modified form, is gratefully acknowledged:

Blackwell Scientific Publications Ltd (Figures 2.3, 4.4, 6.15, 10.3; Table 4.2)
Conseil International pour l'exploration de la mer (Figures 4.7-4.11; Plate 4)
The Controller of Her Majesty's Stationery Office (Figures 2.1, 3.1F, 11.1; Table 5.1; Plate 10)
Dr W. Junk N.V. (Figure 6.13)
Nature (Figure 6.16)
Oikos (Figures 1.2, 1.3, 3.5, 3.6)
Oliver and Boyd (Figure 2.10)
The Royal Society (Figures 2.7, 10.1)
The Society for Water Treatment and Examination (Figure 11.2; Table 11.1)
Thunderbird Enterprises Ltd (Figure 5.5)
E. Schweizerbart' sche Verlagsbuchhandlung (Figures 3.4, 4.13, 4.14, 9.1)

Chapter 1
The Physical and Chemical Background

Temperature

Very practical considerations have led to the development of some knowledge of the temperature tolerances of aquatic animals, plants and micro-organisms. For example, it is known that the trout, a 'cold water' fish, will die at 25 °C and its eggs will not hatch at a higher temperature than 14 °C, while the carp, essentially a 'warm water' fish, can withstand temperatures of up to 38 °C. Also, plant distributions sometimes indicate an environmental change; one example is the water weed *Vallisneria spiralis*, which was introduced into Britain for warm water aquaria and is now naturalized in some places where heated effluents enter rivers.

It is quite customary to speak of an 'optimum growth temperature' when a microscopic plant such as an alga, or a micro-organism such as an aquatic fungus or bacterium, is under consideration. The optimum growth temperature can readily be determined if the organism can be grown in isolation. However, an optimum determined in such a way can be misleadingly high in comparison with the situation in nature. For example, although freshly collected natural material of aquatic fungi is best retained in a healthy condition at a temperature of 10 °C, a pure bacteria-free isolation made from it will often have an optimum growth temperature of 25°C or more. One is here seeing the consequences of the removal of competition, allowing the organism to realize its maximum potential in accelerating the biological process of growth with rising temperature.

Winter conditions in Europe frequently result in the formation of a covering of ice over ponds and lakes. These water bodies rarely freeze solid because of the density properties of water. As the temperature falls down to 4 °C, water density increases. However, as the temperature falls below 4 °C, down to freezing point, water density then falls slightly. The surface water, cooled below 4 °C by a falling air temperature, will be lighter than the water below and will remain at the surface. Therefore, ice formation at 0 °C will occur first at the surface. Once ice has formed the water below will be insulated and will retain a winter temperature of 0·4 to 2·0 °C. This insulation of much of the water body from freezing is a fortunate circumstance for the survival of the flora and fauna, since ice formation disrupts living cells and usually kills them. Growth of algae in fresh water can proceed at very low temperatures and it is not unknown for them to begin their spring maximum increase (see Chapter 2) even before the ice cover has melted away.

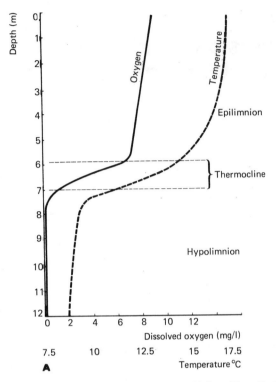

Figure 1.1 **A** Summer temperature and oxygen stratification in Blelham Tarn **B, C** Successive stages in its destruction showing how the thermocline is lowered during this process (from data of Collins and Willoughby)

The differing density characteristics of water at different temperatures can also have important repercussions in the summer season. The warming of the surface of a lake, resulting in water of steadily decreasing density, can proceed to such an extent that a pronounced layering develops. Assuming that the weather is calm, and there is little mixing caused by high winds, the condition of stratification stabilizes to produce a water body which has three distinct horizontal zones. The warm water of the epilimnion circulates within itself but does not mix in with the water of the thermocline, the zone in which there is a very sudden temperature fall within a short vertical distance (see Figure 1.1 **A**). Below the thermocline, the hypolimnion water is of uniformly low temperature and it too circulates within itself but does not rise upwards. The thermocline thus constitutes an effective barrier between the epilimnion and the hypolimnion. A pond rather than a lake will generally not be of sufficient depth for a fully stratified condition to develop, although some kind of temperature discontinuity will often be encountered in the summer. Summer lake-stratification eventually breaks down with the cooling of the surface layers in the autumn and the onset of high winds. Vigorous mixing extends deeper and deeper into the lake and the position of the thermocline falls vertically. Eventually there is no thermocline and the winter condition of tempera-

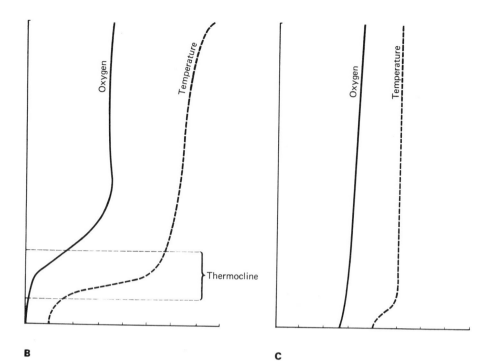

ture uniformity throughout the lake water is re-established. 'Overturn' of the lake has then occurred. The data reproduced in Figure 1.1 (B,C) were in fact obtained from an experimental overturn carried out in the summer by mixing the water of a small lake with powerful pumps[1], but they give a good picture of the pattern of events which occurs naturally in the autumn.

Data from Lake Esrom, Denmark, shows the natural seasonal cycle of temperature experienced in two different types of year (Figure 1.2). Type A includes a warm, calm summer and Type B a cool, windy summer. It will be seen that in Type A the stratified condition is well developed and persists for five months. Overturn and the establishment of uniform temperature conditions occurs late, in October. In Type B the stratification is less well developed and breaks down earlier, in September. These seemingly small differences between year types are important for the animals living in the bottom mud (profundal benthos) in the deepest parts of the lake. Their growth is always slow, because of the low temperature at 20 m depth, but paradoxically they experience more favourable temperature conditions in a year of Type B than one of Type A. In a year of Type B the period of slightly elevated temperature in the benthos before winter sets in is of greater extent and longer duration than it is in a year of Type A.[2]

Running parallel with the summer temperature stratification in lakes, and reinforcing its biological significance, is usually a dissolved oxygen stratification also — this is now discussed further.

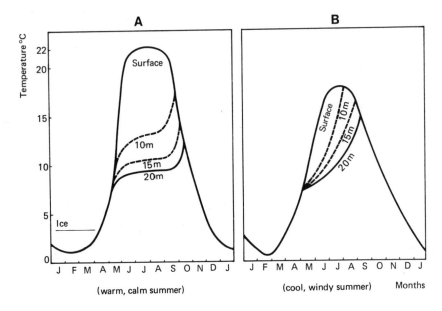

Figure 1.2 Summer temperature stratification and its autumnal destruction in two different types of year in Lake Esrom. Summer temperatures are shown at the surface and at 10, 15 and 20 m depths, the latter being the lake bottom (after Jónasson, 1972)

Dissolved Oxygen

The solubility of oxygen from air, at normal atmospheric pressure, in pure fresh water is related to the temperature of the water by the equation $C_S = 475/(33.5 + t)$, where C_S is the solubility of oxygen in mg/l and t is the temperature. Thus at 6 °C, for example, C_S is 12 mg/l, whereas at 29 °C, C_S is only 7·6 mg/l.[3] The biological significance of this differing oxygen solubility, which is dependent on water temperature, is that plants and animals which respire aerobically may come under stress in warmer water. In handling fish for tagging or pond transfers, it is often advantageous from this point of view to add ice to the baths in which they are being manipulated. This is beneficial not only in making more oxygen available to them but also in lowering their general metabolic activity and prevents them from becoming unduly stressed.

We have been discussing the saturation values for water in contact with air, which contains only 20·9 per cent of oxygen by volume. The saturation values for water in contact with pure oxygen are nearly five times as great. In swiftly flowing waters in upland streams, if supersaturation occurs — caused by the activity of photosynthetic plants or by a rise in temperature — then oxygen is lost to the atmosphere by the ready exchange which takes place. In a sluggish river, with less turbulence, however, oxygen is not exchanged so readily with air. Under these conditions the oxygen produced by photosynthetic plants may be retained and values of up to 200 per cent saturation have been recorded. There is often a diurnal pattern in such supersaturation, with high values during the day and much lower ones during the night, or even an oxygen deficit as photosynthesis ceases and respiratory processes continue. Knowledge of the phenomenon of supersaturation is again useful in handling fish for transport in large tanks over long distances. It is now customary to bubble oxygen, from cylinders of the compressed gas, through the tanks on the journey.

Conditions of oxygen deficit may develop in rivers following the discharge of sewage or other effluents and the biological implications of this are considered in subsequent chapters. Ponds and other small water bodies may also become deficient in oxygen in the summer, particularly if they are overloaded with materials such as decaying leaves. It is then that animals such as *Culex* and *Eristalis* larvae which erect breathing tubes to contact the surface are at a selective advantage. When one considers the larger water bodies, the development of summer temperature stratification in lakes (discussed previously) often results in a marked reduction or even complete elimination of dissolved oxygen at the lower depths. This occurs because the water of the hypolimnion becomes physically isolated and cannot re-aerate at the lake surface. With the continued respiratory activity of the benthic animals and micro-organisms the oxygen store in the hypolimnion falls. Eventually, as in Blelham Tarn (Figure 1.1 A) oxygen at the lower depths may be entirely eliminated. Survival of the invertebrates of the profundal benthos then depends on their endurance of such conditions until oxygen returns in the autumn, when the lake mixes and re-aerates. The extent of the summer de-oxygenation is very much a function of the depth of the lake in question. Blelham Tarn (12 m depth) and Lake Esrom in Denmark (20 m depth) experience total de-oxygenation in the hypolimnion, while in Windermere (60 m depth) a considerable amount of oxygen remains throughout the summer. The reason for this difference is that the larger

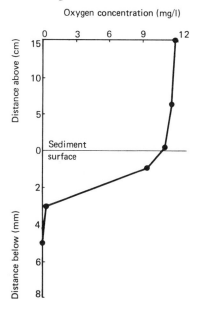

Figure 1.3 Vertical changes of oxygen content above and below the sediment surface of a lake in the winter condition (after Jónasson, 1972 from the data of Hargrave)

volume of the hypolimnion water in Windermere holds a proportionately larger quantity of dissolved oxygen which is not totally consumed in the summer season.

It is as well to bear in mind that even when the water of a lake or pond is thoroughly aerated, the bottom mud (if undisturbed) will only have oxygen in its very surface layers. In Figure 1.3 it will be seen that in an experiment the oxygen was virtually extinguished only 3 mm below the water-mud interface. The natural situation may be more favourable than this as there is always some degree of deep turbulence, but oxygen deficiency is a perpetual problem for the benthic invertebrates.

Calcium

The major cations dissolved in natural fresh waters are calcium, magnesium, sodium and potassium, and of these four calcium and sodium are normally dominant. The presence of calcium is dependent on chemical reactions, involving carbon dioxide, which occur in the following way. Carbon dioxide is a highly soluble gas and enters water directly from rain or more indirectly from the respiratory processes of animals, plants and micro-organisms. As drainage water percolates through soil en route to a stream, river or lake, it gains dissolved carbon dioxide, which combines with water to form carbonic acid, H_2CO_3. Although this acid is weak, it is nevertheless capable of acting on carbonate-holding rocks and releasing a soluble product; from limestone rocks calcium bicarbonate, $Ca(HCO_3)_2$, is produced. The presence of a generous amount of dissolved calcium bicarbonate renders a water 'hard', with a high alkalinity (pH 7 to 9). Such a water will generally facilitate the production of both plants and animals, and here the biological significance of calcium is made plain. On the other hand, if drainage water passes over non-calcareous rocks such as granite, shales or slates, very little calcium will be present — it will be 'soft' and may be highly acid. A soft water is generally much less productive than a comparable hard one. The reason for the difference in productivity between hard and soft waters is not fully established but it seems possible that the more efficient re-cycling of nutrients by microbes (bacteria and fungi) in the former is a relevant factor. The acidity of waters of bogs and bog lakes, which are extremely unproductive and may have a pH as low as 4, is often partly attributed to the precipitation of sulphuric acid from the atmosphere through rain; this effect is particularly pronounced if the situation is close to a site of heavy industry. The formation of highly acid brown materials, humic acids, in the waters themselves is also contributory. Why the acidity of atmospheric derivation does not affect hard waters also may be questioned, and here one can note an important benefit which stems from the presence of calcium bicarbonate in that it acts as a buffer to neutralize this acidity. If sulphuric acid is added to a hard water, the following reaction takes place

$$Ca(HCO_3)_2 + H_2SO_4 \rightarrow CaSO_4 + H_2O + CO_2$$

The strong acid, sulphuric acid, is replaced by an equivalent amount of carbonic acid, a weak acid, and the pH does not fall steeply.

When one considers the various types of freshwater situation from the point of view of the presence of carbon dioxide, rather than that of calcium, the following picture emerges. In soft waters, with a low pH, there is little carbon dioxide bound as bicarbonate or carbonate and the gas is found to be simply dissolved in the water. Harder waters on the other hand contain carbon dioxide bound as bicarbonate in addition to some dissolved and free. Finally, in extremely hard waters there is no free carbon dioxide and the equilibrium reactions move towards the production of calcium (or magnesium) carbonate which, because of its low solubility, tends to be precipitated. These relationships are depicted in Figure 1.4. Generally photosynthetic plants can utilize either free carbon dioxide or that bound in bicarbonate for their carbon fixation, but the aquatic moss *Fontinalis* is an exception in that it cannot utilize the latter.

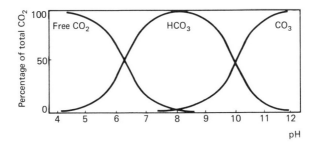

Figure 1.4 The relationship of pH to the percentage of total carbon dioxide in each of its forms in water

A complication in examining published data for water hardness is that these are frequently expressed as mg of $CaCO_3/l$ rather than as bicarbonate, which is actually estimated. A calculation converts one value to the other, and we can only wonder at the sometimes devious ways of chemists.

Animals, such as snails, which deposit calcium carbonate in their shells have a very obvious requirement for calcium, and generally do better in hard waters than in soft ones. However, the fact remains that calcium extraction for shell formation can proceed in very dilute solutions — the situation regarding the freshwater shrimp *Gammarus* is discussed in Chapter 5.

Nitrates and Phosphates

Nitrogen and phosphorus are basic elements in all living matter and it is not unexpected to find that the water-soluble forms of these, nitrates and phosphates, are important regulators of biological growth in fresh water. The free-floating algae, and especially the small planktonic forms, are affected directly, and high dissolved nitrate and phosphate levels will generally stimulate their growth. This always assumes that other accessory growth substances such as iron or silicon (for diatoms) are present in sufficient quantity as well. The complication in regard to nitrates and phosphates is that, in addition to the significant amounts of these chemicals which are re-cycled within the freshwater environment, others can enter from outside sources. Water bodies situated in areas of fertile land will gain these growth nutrients from soil leaching and from the type of effluent produced by farms, especially if intensive livestock breeding is practised in them. Human sewage effluent, even if it is treated in a sewage works before discharge, will also give a sizeable nitrate and phosphate addition to the water which receives it. The effect of these additions from extraneous sources can be measured in terms of the algal productivity. For example, in nutrient-rich lakes in Denmark and northern Germany the planktonic algae produce up to 2400 g of organic matter per square metre of surface per annum, while in similar lakes with a more sterile drainage area the production is only 100 g.

References

1. V.G. COLLINS and L.G. WILLOUGHBY, 1962. *Arch. Mikrobiol.* **43**, 294-307.
2. P.M. JÓNASSON, 1972. *Oikos,* Supplementum 14, 1-148.
3. F.J.H. MACKERETH, 1973. Freshwater Biological Association, Scientific Publication No. 21.

Chapter 2
Photosynthetic Plants

Photosynthetic plants occurring in fresh water comprise algae and vascular plants, and it is usually convenient to consider them separately. However, there are some common requirements for all these plants, namely adequate light and carbon dioxide for energy fixation processes, oxygen for respiration, and supplies of major elements such as calcium, nitrogen, phosphorus, potassium and sulphur. In addition, smaller amounts of other elements such as copper, iron, manganese and zinc are necessary, and trace amounts of organic growth factors such as vitamins may even be required. Certain groups of the photosynthetic plants have requirements which fall outside this basic pattern; for example, the diatom algae need silicon as a major element in order to construct their cell walls.

The pigmented regions of photosynthetic plants are involved in the conversion of light energy to chemical energy, and the composition of the pigments and their role in photosynthesis varies in the different groups. Whereas in the vascular plants the important photosynthetic pigments are chlorophylls-a and -b, both absorbing strongly in the red (650-680nm) and blue (400-450nm) regions of the spectrum, the freshwater algae tend to utilize chlorophyll-a and possibly a smaller amount of chlorophyll-b in addition, and a variety of other pigments as well. For example β - carotene is a major photosynthetic pigment in both the diatoms and the dinoflagellates (for example, *Ceratium*) and it tends to modify the colour of the photosynthetic region to orange or brown rather than the familiar green of the vascular plants. The significance of carotenoid pigments such as carotene in photosynthesis seems to lie in their capacity to absorb light energy and transfer it to chlorophyll-a They are effective in the blue and green regions of the spectrum (430-500nm). The basic equation of photosynthesis is as follows

$$CO_2 + 2H_2O \longrightarrow (CH_2O) + O_2 + H_2O$$
$$\downarrow$$
$$C_6H_{12}O_6$$
$$\text{Glucose}$$

In considering the light requirement, the expression 'the photic zone', essentially the depth to which plants can obtain sufficient light to grow, is in widespread use. Obviously the extent of the photic zone will depend on many local factors such as the amount of incident light, the turbidity of the water, the possible discolouration by dissolved humic substances, and so on. Under normal circumstances there is sufficient light penetration into lakes to allow the growth of planktonic algae down to about a depth of 10 m, or even lower (see reference to Lake Superior on page 27) Benthic algae inhabiting the bottoms of ponds, lakes and rivers show a definite distribution pattern based on light penetration and do not normally occur in a vigorous condition at depths greater than about 6-8 m. Water readily absorbs wavelengths in the infra-red and ultra-violet regions of the spectrum; for example, ultra-violet is reduced to 5 per cent of its incident value at only two metres depth, but the red, blue and particularly the green wavelengths penetrate much deeper. Water in a particularly static condition may show a depression of algal growth at the extreme surface due to the sterilizing effect of the absorbed ultra-violet. Utilization of the red and green wavelengths by the phytoplankton as it photosynthesizes causes some attenuation of these wavelengths, and if growth is very intense a 'self shading' effect may even be apparent. Thus the spring growth of *Asterionella* in Windermere may be so prolific (see below) that it actually affects the pattern of light penetration into the water. This effect is greatest for the light rays in the green wavelengths; their depth of penetration falls by about a third at such times. While 5 per cent of the incident amount occurs at 10 m in normal times, this value is recorded at only 7 m during such a spring growth period. Light rays in the red and blue wavelengths are affected less, and those in the ultra-violet, which have a much shallower penetration anyway, are hardly affected at all. Self-shading effects by *Asterionella* are only stable for a few weeks in Windermere, in the month of May, and quickly disappear when shortage of nutrients and turbulence leads to dispersal of the algal mass.

Light requirements of the submerged vascular plants growing in fresh water parallel those of the algae. A convenient yardstick is that these plants, which are rooted, may colonize substrates down to depths where the light intensity is 1 to 4 per cent of that at the surface. With few exceptions this means that these vascular plants, and others included with them in the category of hydrophytes (see below), are confined in lakes to the upper ten metres of water; but the exceptions are of interest.

Fontinalis is recorded down to 20 m in exceptionally clear lakes and *Nitella* is reported to grow at 27 m depth in lakes in Japan. In a survey, which is considered further on, of the Scottish lochs, Spence found that the greatest depths for hydrophyte colonization were in clear calcareous waters. Shallower limits existed in waters containing brown humic substances and even more shallow ones were present in chemically rich waters bearing an abundant plankton.

The carbon dioxide requirement of photosynthetic plants is generally satisfied by an abundant supply, either as the free gas or as bicarbonate ions in the bottom muds or their overlying waters. Photosynthesis by the phytoplankton in lakes may be so intense that the carbon dioxide content of the surface water is temporarily reduced, leading to a rise in the pH of the water. The pH of the epilimnion of Esthwaite water has risen from 7 to 9 during such times.

Because photosynthetic plants generate oxygen as an end product of photosynthesis, they have an important role in aeration processes in fresh water, especially in rivers. Not all the oxygen can be retained in the water immediately, and up to 7 per cent of the daily production may be lost to the atmosphere in bubbles escaping from plant surfaces. However, during the hours of sunlight, plant communities may produce a dissolved oxygen concentration in the water which is in excess of that normally dissolved from air, and values of 200 per cent saturation have been recorded. This phenomenon of supersaturation is usually transitory however, and oxygen is lost to the atmosphere as diffusion gradients stabilize, or even more rapidly if the water is agitated by passage over waterfalls or weirs. In both standing and flowing waters where plants are abundant the dissolved oxygen regime may show a pronounced diurnal rhythm, which is a reflection of a net oxygen production by day and a net oxygen demand by night, when respiration continues in the absence of photosynthesis (Figure 2.1). In such situations the nocturnal fall in dissolved oxygen may place the fauna in jeopardy and a very misleading picture would be presented by a biologist who sampled only in orthodox working hours!

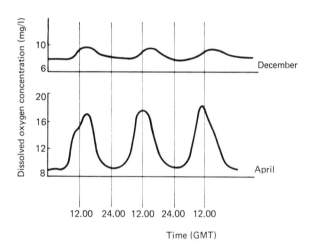

Figure 2.1 Diurnal rhythm of dissolved oxygen in an English river. Daytime increase of oxygen in December attributed to submerged vascular plants and in April to the growth of epiphytic diatoms as well (adapted from DSIR 1960)

Algae

The basic pattern for the metabolism of a freshwater alga is the purely autotrophic one, involving the fixation of carbon dioxide by photosynthesis and the elaboration of cell constituents from chemically simple starting materials. Assuming a maximum efficiency in the metabolic processes, carbon dioxide derived from respiration would be the sole waste product of the carbon assimilation. However, these two features of the basic pattern do not always apply. On the one hand many algae can utilize organic substrates such as sugars or organic acids and by so doing live in complete darkness, while on the other hand many algae excrete organic compounds into the water in which they grow. The heterotrophic mode of life of certain algae—their predilection for organic growth substrates—has been demonstrated in nature when plankton grew in sub-arctic lakes in the winter when there was virtually no light available. It was assumed that organic material produced by photosynthetic activity in the summer was being utilized. The heterotrophic mode of life of an alga may be so extreme that photosynthesis will not occur even if light is present. However, this is especially true of forms from heavily contaminated and polluted waters and no strictly planktonic form is in this category. Laboratory studies with the planktonic strain of *Chlorella pyrenoidosa* show that this cannot grow in complete darkness, but in limiting light intensities enhanced growth can result from the provision of a glucose supply. Therefore, it is possible that this alga may utilize soluble organic materials which occur in nature. It is to be expected that if organic materials such as glucose, polysaccharide, or acetate do become available naturally then the planktonic algae will compete for them poorly, in competition with the bacteria which are always present. Although this expectation is usually realized there are conditions when the algae can compete successfully. These occur when some light is present; by using photochemically generated adenosine triphosphate (ATP) a very active uptake mechanism is set in motion. This photoassimilation of organic materials by algae often occurs at limiting light intensities, for example, at the bottom of the photic zone in lakes.

The organic compounds (extra-cellular products) excreted by algae into freshwater, sometimes comprise a surprisingly large proportion of the carbon which is fixed in sunlight; one study suggested that algae at a lake surface on bright days excreted no less than 95 per cent of their fixed carbon. Of the various extra-cellular products of freshwater algae, glycollic acid, $CH_2 OH \cdot COOH$, has been studied most and much is known about it. Glycollic acid is derived from one of the sugar diphosphates of the carbon-fixation cycle, possibly even from ribulose diphosphate, the immediate acceptor of carbon dioxide. Hence its production is closely linked to the carbon-fixation or photosynthetic cycle. Glycollic acid is not a breakdown product produced by dying cells – using ^{14}C tracers it has been found to be one of the few organic compounds to appear in culture solutions containing young and vigorously growing cells. When an alga is grown in culture, the amount of glycollate excreted depends on the environmental conditions; for example, the greatest amount is excreted when cells are transferred from low light and high carbon dioxide concentrations to high light and limiting carbon dioxide concentrations. Evidently the cell membranes are readily permeable to glycollic acid and in a culture flask the amount excreted represents an equilibrium between intra and extracellular concentrations. It has been argued that glycollic acid must build up to a certain level before growth can begin; this statement is based largely on the observation that addition of a small amount (1 mg/l) shortens the lag phase. If this is the case in nature, then glycollic acid production and liberation may be very important, but the field evidence is that only very minute levels can be detected in lake water, namely, 0·06 mg/l or even less.[2]

There is also the possibility that glycollic acid excreted by algae at one season of the year may be utilized by them at another: they would be in the strongest position to do this by employing a photoassimilation mechanism as mentioned above. Since many bacteria can use glycollate readily as an energy source, there is the possibility that this material may actually be the attractant which encourages bacteria to inhabit their surfaces. The association could be beneficial to the alga; this aspect will now be discussed further.

Phytoplankton

Planktonic organisms are those which are freely suspended in water and are possibly aided in this by limited ability to move. However, since the algae constituting the phytoplankton are rarely motile, they rely on buoyancy to retain a high position in the water, thereby ensuring that photosynthesis can take place. Many planktonic algae have elaborated spines or long processes, or they may grow as colonies in which the individual cells radiate from a common centre. *Asterionella* (Figure 2.2 A) exemplifies the latter mode of growth. These individual cell elaborations or colony aggregations are generally thought to be devices to increase buoyancy by conferring a large surface-area to volume ratio, but their buoyancy is greatly dependent on natural turbulence in the water. *Asterionella* colonies sink when removed from their natural environment into perfectly static water. Our views on cell elaboration as a buoyancy aid have been under further re-examination following the recent discovery that many planktonic algae have mucilage coverings which may completely clothe the cell processes. Clearly, the situation, like so many others in biology, is more complicated than it would appear at first sight. The presence of gas vacuoles in algal cells may also contribute to their buoyancy; this is especially so in the blue-green algae and is discussed later.

A well-known characteristic of phytoplankton in temperate waters is its greatest development in the spring season, resulting in the so-called 'spring maximum'. This occurs in lakes all over Europe and is also reported from North America and New Zealand. The spring maximum is a result of the combination of several factors. There are seasonal fluctuations in the concentrations of chemical nutrients in the water, but in mid-winter they are at maximum levels. Since algal growth is so dependent on these nutrients, the stage would appear to be set for vigorous exploitation of the favourable chemical condition at this time. The numbers remain low, however, although exceptional species such as *Melosira italica* (see page 25) can take advantage of the conditions. One unfavourable factor is the incidence of rough weather at this time of the year; this continually mixes the whole water body to such an extent that the phytoplankton is often carried down from the limited depth in which it can photosynthesize (the photic zone). If mixing nullifies the multiplication which can occur in the photic zone, then no net increase of cells results. In Windermere we may take it that 10 m is the maximum depth of the photic zone, as compared with the maximum depth of 64 m recorded for this lake. The potentiality of lake water collected in the winter time to support phytoplankton growth is seen when Windermere samples are returned to the laboratory and illuminated there. Growth of *Asterionella* results. Another factor which prevents the build-up of many species of planktonic algae in lakes during the winter months

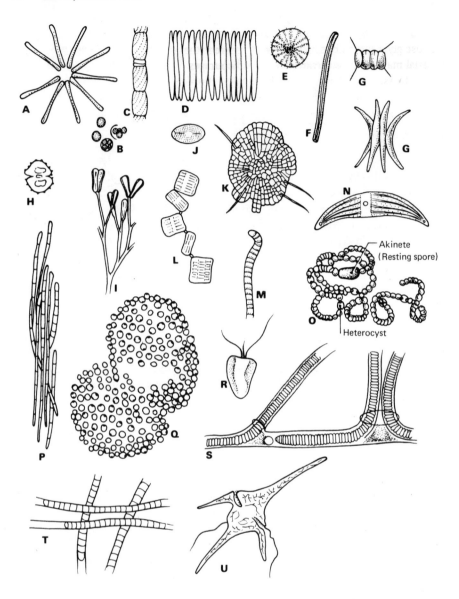

Figure 2.2 Freshwater algae A *Asterionella* B *Chlorella* C *Melosira italica* D *Fragilaria crotonensis* E *Stephanodiscus niagarae* F *Nitzschia* G *Scenedesmus* H *Cosmarium* I *Gomphonema* J *Cocconeis* K *Coleochaete* L *Tabellaria flocculosa* M *Oscillatoria* N *Closterium* O *Anabaena circinalis* P *Aphanizomenon* Q *Microcystis aeruginosa* colony R *Prymnesium parvum* S *Plectonema* T *Lyngbya* U *Ceratium* A C−F I J L are diatoms: Bacillariophyceae; B G H N K are green algae: Chlorophyceae; M O−Q, S T are blue-green algae: Myxophyceae/Cyanophyceae; R is a flagellate, U is an armoured flagellate: Dinophyceae. (All highly magnified, drawn to various scales.)

is the large loss of cells constantly occurring through the outflow streams and rivers. This loss is a reflection of the total rainfall and the wetter is the winter the lower is the phytoplankton population when spring arrives.

The phytoplankton may be reduced to such a low level in the winter that germination of algal resting stages has been suggested as the trigger for the spring maximum development. However, no such stage is known for the important planktonic diatom *Asterionella* and it appears that some living cells of this alga are always present either in the main body of the lake, or in the sheltered bays, or even arriving in the waters of the inflowing streams. With a general rise in the temperature of the water, the increasing stability of the water body and the declining loss of cells via the outflows, the spring increase of the phytoplankton begins. In Windermere the spring increase of *Asterionella*, occurring usually in the months of March and April and reaching a maximum in May or even June, has been followed in relation to the availability of essential nutrients which the alga must have to grow and divide. The silicon relationship is particularly interesting. *Asterionella* builds silicon into its shell, the weight of which is independent of environmental conditions. It follows that with the rapid spring increase of *Asterionella*, dissolved silicon in the water is consumed rapidly; its replenishment can only occur through the inflow rivers and streams since its initial derivation is the surrounding soils. In Windermere the reservoir of dissolved silicon present during the winter is 1 mg/l or slighty less, and this store does not decline initially as *Asterionella* begins to grow but eventually does so, when a large population of the alga builds up. By the time 10^6 cells/l are present, lake-water silicon is falling rapidly, and with a drop to about 0·25 mg/l no further divisions of the alga can take place (Figure 2.3).[3] Phosphorus in the water occurring in the form of phospate (the most available form to algae) also declines during this time as the result of the algal growth, and in fact its pattern of decline slightly precedes that of silicon, because algal cells consume large quantities of phospate if it is available and store it for subsequent cell divisions. The decline of phosphorus, as phosphate in the water, during the *Asterionella* growth period is from 0·003 mg/l in March to 0·0004 mg/l or less at the end of May. The rate of increase of *Asterionella* in the spring is measured by the cell 'doubling time', and this is estimated at 5 to 7 days; it falls far short of the 'doubling time' of ten hours achieved in the laboratory and indicates that the growth in nature is proceeding in conditions of temperature and light which are far from ideal, despite the astronomical increase in numbers.

Following the cessation of the massive phytoplankton growth in the spring in Windermere and other similar lakes, there is an inevitable decline in the populations as many of the cells age, die, and eventually find their way to the bottom mud of the lake (Figure 2.3). However, some living cells persist in the plankton and a low level of growth is maintained during the summer. During this time the thermal stratification (see Chapter 1) developed in the lake tends to isolate the surface waters from those beneath, thus preventing any replenishment of nutrients into the photic zone from the lower waters and bottom muds. However, further supplies can still arrive through the inflow rivers and streams. Nutrient replenishment from the lower waters (hypolimnion) eventually does

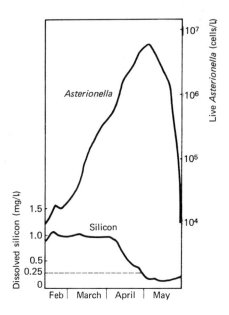

Figure 2.3 The growth of *Asterionella* in Windermere, South Basin, in relation to dissolved silicon (after Lund, 1950)

occur when thermal stratification breaks down in the autumn and a second spurt of phytoplankton growth, known as the 'autumn maximum' then ensues. This produces far fewer algal cells than does its spring counterpart, not because nutrients again become limiting as they did in the spring, but because incident light is diminishing as the season advances and with the advent of rough weather the cells have only limited time in the photic zone before being mixed back into the main body of the lake. Because complete anaerobiosis of the bottom waters and muds occurs during the summer in some lakes (e.g. Esthwaite Water), but not in others (e.g. Windermere), it has been argued that the autumnal maximum of phytoplankton growth should have a different magnitude in the two types. This argument is based on the knowledge that anaerobiosis of the bottom muds leads to their releasing phosphate, but this is not released and is bound to iron and other cations if the mud is aerobic. Consequently the breakdown of thermal stratification would be expected to release more phosphate to the phytoplankton if the lake had been anaerobic during the summer than it would if the lake had not; hence there would be expectations of a larger autumn maximum of phytoplankton in Esthwaite Water than in Windermere. These expectations are not realized to any obvious degree, and speculation continues on the full extent and implication of phosphate release from lake bottoms in the autumn.

Grazing by the invertebrate animals which occur in lake water does not have any appreciable effect on the spring increase of *Asterionella*; on the other hand there is evidence that the eventual deposition of the cells of this and other large algae into the surface benthic mud has a stimulative effect on the animals there, which use them as food. In this connection, work in Danish lakes has shown that the two most important periods of growth of the common midge (*Chironomus anthracinus*) larva are the spring and autumn (see Chapter 3).

Although much is known, from its ecology, about the nutritional requirements of *Asterionella*, and it grows well in the laboratory in rough culture, it has so far proved impossible to cultivate it in a completely pure, or axenic (bacteria-free) culture. It follows that the growth requirements cannot be defined completely and the idea has developed that this and certain other planktonic algae in fresh water may have special syntrophic associations with bacteria which occur on their surfaces. In light of the knowledge that many algae of the marine phytoplankton require the vitamins B_1 and B_{12} to grow, it has been suggested that these or other more exotic requirements in certain freshwater algae might be satisfied by their bacterial partners. Glycollic acid is another substance which has been mentioned in this connection. However, this issue is unresolved at present.

In addition to the algae of the phytoplankton such as *Asterionella*, which show regular spring and autumnal pulses of growth clearly related to environmental conditions, others show more complicated patterns. An example is the filamentous diatom *Melosira italica* (Figure 2.2 C), which can make rapid spurts of growth and is then capable of achieving a doubling time which is three to five times that of natural *Asterionella* populations. This alga seems to prefer cooler conditions for its growth and tolerates low light intensities. In the English Lakes it is most abundant in the autumn, winter and late spring and decreases as water turbulence falls away in the summer. In the latter season its eventual dramatic disappearance in the plankton has led to a search for the viable cells which eventually restore the population. It appears that in this instance special resistant cells survive anaerobic conditions in the bottom mud of the lake and resume growth on being re-suspended in the water when favourable conditions return. The bulky nature of this alga makes it very dependent on water turbulence to maintain a position for it in the photic zone.

Lakes in tropical climates have a temperature and light regime which is suitable for phytoplankton growth throughout the year — in Lake Victoria in Africa a moderately high productivity is ensured in all seasons by efficient mixing of nutrients mediated by wind action in this extensive but shallow lake basin.

Lake George, Uganda, typifies a rather special kind of equatorial lake where diatoms are of little importance and blue-green algae are dominant. They apparently receive sufficient nutrients to grow continuously and the temperature regime is always favourable, ranging from 25°C to 35°C during the year. The reason for the particular success of the blue-green algae in these lakes is not entirely clear but their capacity to fix atmospheric nitrogen (see page 32) may play a part in this. Another possible factor is the known ability of the blue-green algae, rather than the diatoms or green algae (Chlorophyceae), to survive conditions of darkness and oxygen lack to a marked degree. Lake George has a depth of 2 m only, and *Microcystis aeruginosa* (Figure 2.2 Q), the most common blue-green alga, tends to sink to the bottom mud during the hours of daylight. Although there is some re-suspension in the water column when increased turbulence occurs at night, this is generally not sufficient to bring the whole population back into a suitable environment for growth. However, when strong winds occur the bottom sediment is stirred and the deposited *Microcystis* is then re-suspended and resuscitated to grow again.[4]

Comparative Phytoplankton Studies

From the foregoing discussions it is plain that the soluble substances which occur in water are very influential in determining the amount of phytoplankton growth which takes place. When different lakes are compared, those with waters deficient in major elements such as nitrogen or phosphorus generally have a smaller phytoplankton than do chemically enriched lakes. Another aspect of enrichment which touches on the phytoplankton is that the species of algae present in any single lake may have changed in the course of time, to accommodate for presumed changes in the water chemistry (see the discussion of *Asterionella* in Chapter 10). Present-day comparative studies of adjacent lakes which differ in one or more major environmental factors have as their objective the evaluation of the causal factors affecting phytoplankton growth.

In large scale comparative studies on the growth of algae, the labour involved in examining, counting and analysing the populations may be so enormous that more 'shorthand' methods are sometimes adopted. One of these involves the extraction and quantitative estimation of chlorophyll in the water samples. Another method measures the assimilation of ^{14}C made by the algal pigments as they photosynthesize, in water samples enclosed in bottles. In this method, the water samples, containing the algae, are recovered from the particular depth under investigation, ^{14}C is added as $NaH^{14}CO_3$, and the samples are then often returned once more to the same depth. After a suitable exposure time, the samples are filtered and dried and the ^{14}C incorporated into the algae is estimated using a Geiger counter. In addition to the photosynthetic assimilation of the carbon, a 'dark fixation' may also occur (see above) and a correction is made for this by exposing similar bottles with a black-painted exterior. The 'dark fixation' of carbon revealed by this latter exposure does not normally exceed 2 per cent of the photosynthetic assimilation. Two comparative studies where the 'shorthand' methods were used will now be discussed.

Lakes near centres of human population are usually subject to such a variety of influences that the effect of one of these cannot be easily isolated from the effect of the others. In a recent study in the Canadian high-arctic, however, the phytoplankton in two adjacent small lakes was examined in light of the knowledge that one, Lake Char, was perfectly natural, while the other, Lake Meretta, received sewage and waste water from a government camp with a fluctuating population of fifty to a hundred inhabitants.[5] Both lakes had low water temperatures of less than 5 °C throughout the year and froze to a depth of two metres in the winter. This was during the period of the polar night from early November until early February.

The summer period of continuous light extended from early May until mid August. Both lakes were completely free of ice only from mid August until mid September. Samples for phytoplankton analysis were not examined microscopically for species composition, but an estimate of activity was made by estimating chlorophyll-a present in a unit volume of water. After filtration through glass fibre, the phytoplankton was extracted with acetone and assayed in a fluorometer calibrated with a spectrophotometer. In Lake Char chlorophyll-a values ranged from 0.13 mg/m^3 at the depth of the winter to 0.69 mg/m^3 in May, with no suggestion from the parallel cell count that there was any appreciable heterotrophic growth in the former period. These maximum and minimum values agreed closely over several years. In Lake Meretta there was a dramatic effect attributed to the addition of the sewage. This flowed in during the summer but remained frozen below the outfall during the winter. Chlorophyll-a values were from four to eight times those from Lake Char in the growing season and ranged from 3 to 5 mg/m^3. A July pigment peak of 24 mg/l was found in one year only, giving clear evidence that the amount of the nutrient addition and its pattern of utilization was not a consistent annual event.

When adjacent lakes are as enormous as those comprising the North American Great Lakes system, any overall comparison of the growth of algae in them, seen against a background of differing environmental factors, is correspondingly more difficult to accomplish. From Lake Superior in the west to Lake Ontario in the east, the system spans 750 miles and substantial vessels are required for sampling. Lake Superior has the cleanest and clearest water and a photic zone 36 m deep, while in Lake Michigan to the south the inshore photic zone is 12 m only, but extends to 20 m offshore at a distance of five to eight miles from land. However, it has proved possible to compare the overall algal productivities of these two lakes in detail and relate the others to them.[6] Parallel determinations were made using two methods. Uptake of ^{14}C, and the pH change (this bears a relationship to carbon dioxide consumed) shown by water samples exposed in natural light for six hours on board ship, made it possible to derive a yield in terms of carbon fixation. A full analysis of the photic zone profile was made and the result was expressed on a square metre basis. In spite of some discrepancy in the results from the two methods, it was clear that the carbon fixation per square metre was very similar in both inshore and offshore waters of Lake Michigan and was three times that occurring in Lake Superior at the same time of the year. The dominant algae in the open waters of the Great Lakes are diatoms with *Asterionella formosa, Fragilaria crotonensis, Melosira islandica* and *Stephanodiscus niagarae* (Figure 2.2) well represented in Lake Michigan. In addition to these, blue-green algae are an important component of the summer flora, especially in Lake Erie to the south. At its western end, Lake Erie receives important nutrient injections from effluents from the city of Detroit and from Michigan and Ohio farm run-off. Its dissolved phosphorus level is around ·025 mg/l as compared with less than ·01 mg/l in the other Great Lakes. Preliminary work suggests that the carbon fixation per square metre of this lake will not be uniform. At its western end, Lake Erie's carbon fixation value is approximately eight times as much as Lake Michigan's, and decreases to a much lower figure

Figure 2.4 North American Great Lakes compared. Shading density is approximately proportional to carbon dioxide consumption from the water and hence to algal productivity (from data of Verduin)

at its eastern end. This points to the nutrient addition at the western end of Lake Erie as being an algal growth stimulant (Figure 2.4). The North American studies have suggested that growth rates of algae in the Great Lakes are not always conditioned purely by the supply of radiant energy and nutrients. Each species has inherent growth characteristics. For example *Stephanodiscus* has a characteristic slow growth rate of about 30 days to double a population, irrespective of the lake and site. On the other hand *Melosira islandica*, as studied in Lake Michigan, has a slow growth rate in offshore waters (31 days doubling time) as compared with a more rapid growth rate in the inshore waters (19 days doubling time).

Benthic Algae

The simplest definition of a benthic community is that it is an assemblage of organisms on the bottom of ponds, lakes and rivers. However, the benthic algae clearly cannot grow normally on the bottoms of deep lakes where there is no light. With this reservation they occur on the sediment (epipelic forms), on rocks and stones (epilithic forms) and on benthic plants (epiphytic forms) (see Figure 2.5). Whereas a lake of moderate size such as Windermere often contains a single distinct

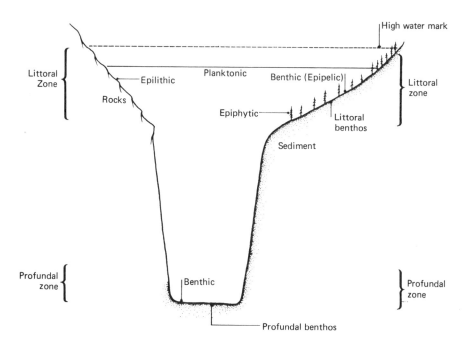

Figure 2.5 Schematic view of a lake, not to scale, with terminology in general use. The littoral zone extends down to the limit of actual or possible colonization by rooted aquatic vegetation. Terms *inside* the water body are descriptive of algal and other biological communities and sub-communities occurring in the situations indicated.

phytoplankton community which varies little from place to place, the benthic algal flora will be developed to varying extents in different regions of a lake. These floras are affected by the differing substrates which are available, by the inflows and the outflows, by wind-induced currents, or by depths.[7]

Since the benthic algae are always intimately associated with a substratum, they are very difficult to sample adequately and the result of this difficulty is that knowledge of their distribution and activity lags behind that accumulated for the phytoplankton. However, there are situations where benthic algae may be much more important than the planktonic forms, for example, in running water where the phytoplankton is poorly developed. Again, in ponds the benthic algal flora is often much more developed than the planktonic one.

The epipelic assemblage of benthic algae in lakes is essentially a sub-community of motile forms. The power of movement is a prerequisite for their life on the sediments, which are continually being disturbed by currents and by animals, and it is assumed that this movement is a phototaxis. In line with this assumption is the knowledge that the epipelic diatoms, such as *Nitzschia* (Figure 2.2 F) can be sampled by placing a coverslip on the surface of freshly collected benthic mud taken from shallow water. The algae move up to the underside of the coverslip and can be examined after they have accumulated there for twenty-four hours. There are, however, a few exceptions to the 'rule of motility', for example, *Scenedesmus* (Figure 2.2 **G**) is non-motile. On the sediments of small ponds, especially in moorland situations, a relatively non-motile flora occurs. This flora is made up of mucilage-forming genera and desmids which are both unicellular, for example, *Cosmarium* (Figure 2.2 **H**) and *Micrasterias*, and also filamentous, e.g. *Hyalotheca*. These pond sediments, however, are not greatly disturbed by wave action or animals, as are the sediments of the littoral zone of lakes. The desmid genera mentioned are scarce or lacking in the latter. On the sediments of streams and rivers, with a moderate to fast flow, the flora consists almost exclusively of motile species, and only in conditions of very slow flow can the non-motile species survive. The idea that the epipelic algae of the benthos become detached and give rise to the plankton was held for some years but is now considered erroneous. The idea arose because planktonic forms such as *Asterionella* seemed to disappear from the plankton at certain times of the year and the existence of other reservoirs was accordingly suspected. However, it is known that there is never a build-up of planktonic forms on the sediments before a lake plankton bloom, and in any case the general prerequisite of motility for the epipelic flora means that few truly planktonic species could survive in the sediments.

The epilithic and epiphytic algae of the benthos are essentially non-motile. Some of the diatoms, for example *Cocconeis* (Figure 2.2 **J**) are stuck down with mucilage like postage stamps, while others, for example *Gomphonema* (Figure 2.2 **I**), have attachment stalks. In *Coleochaete* (Figure 2.2 **K**) and *Ulvella*, prostrate discs of cells are formed. The epilithic algae are particularly difficult to observe; recourse has been made to methods such as applying collodion films to stones and later stripping them off after exposure. More recently, special fluorescence microscopy has been used with success. The epiphytic forms are often loosely attached, for example *Tabellaria* (Figure 2.2 **L**) adheres in chains to the submerged stems of *Phragmites* reeds. Within the epilithic and epiphytic sub-communities, particular associations with the substratum are often noted, and these may be very specialized. For example, certain epiphytes grow only in the furrows of leaves.

In addition to a general pattern of distribution based on the chemistry of water— for example, some species of benthic algae prefer alkaline and others acid waters — the chemical nature of the sediments may be important. In a single lake several different types of surface sediment may occur. For example, in Malham Tarn, Yorkshire, both a peat sediment and a coarse calcareous sediment occur, each with its own epipelic algal flora. A black sediment beneath *Chara* beds in this Tarn is almost devoid of an epipelic flora, and this is attributed to the toxic effect of hydrogen sulphide released by decaying organic matter.

Seasonal growth cycles occur in the benthic algae and the overall pattern is similar to that seen in planktonic species, suggesting similar factors operate in both communities. In temperate climates the benthic diatom flora begins to grow strongly in February and March with a peak in April and May; less growth occurs in the summer and a smaller autumnal peak is over by November. In Malham Tarn there are few diatoms in the plankton from mid-winter to mid-summer but the silicon in the water falls from 1 to 0·2 mg/l, as it does in Windermere. This fall is attributed to growth of the benthic diatoms, which are generally well developed in this Tarn.

Addition of organic matter has a discernible effect on the composition of the benthic flora, blue-green algae and flagellates being favoured. Heavy organic pollution of streams and rivers may reduce the flora to a few or even a single species, for example, the blue-green alga *Oscillatoria* (Figure 2.2 M) is particularly noted here. Somewhat surprisingly species of *Closterium* (Figure 2.2 N) are occasionally abundant in such polluted habitats, possibly indicating that this desmid requires an organic growth factor.

When there is grazing of the epipelic benthic algae by ciliates and other small animals, the overall effect is complicated by the nature of this feeding, which can be very selective. In some small ponds, grazing may be so severe that only species resistant to passage through animals survive, for example, in one situation investigated, only *Scenedesmus* was present as a result of this.

Algal Blooms

The sudden appearance of blue-green algae as 'blooms' or 'breaks' at the surfaces of productive lakes has been commented on frequently and is usually taken as a sign that chemical enrichment favouring the growth of algae is at a high level. Bloom formation only occurs when there are high numbers of the algae present in a lake, and the resultant visual spectacle represents the sudden concentration of a hitherto more dispersed population rather than the formation of a new one. Although the event seems more prosaic, less dramatic, in the light of this knowledge, certain facets are full of interest, notably the reasons why the algae may aggregate in such a fashion. A study of bloom formations of *Anabaena circinalis* (Figure 2.2 O) in Crose Mere, Shropshire, indicates that photosynthetic activity and internal vacuole formation are intimately inter-related in such a way as to control the buoyancy of this alga and hence its potentiality to rise and aggregate as a bloom.[8] The internal vacuoles have constructed walls and are gas-filled; the volume of the cell they occupy varies according to the rate of growth of the alga. Thus in the warm period leading up to the time of bloom formation, in May, growth of the alga was rapid and total vacuole volume was 1·1 per cent of the cell volume, with some evidence that the production and erection of gas-vacuoles lagged behind the rate of cell division. Once growth had ceased due to the limitation of nutrients, the gas-vacuole volume rose to 3·8 per cent of the cell volume and the cells became more buoyant. At this stage *Anabaena* was in the epilimnion, down to a depth of 6 m, and was generally very buoyant but was also undergoing regulatory movements in a vertical direction. Thus, when filaments of the alga sank out of the illuminated zone, the cells suffered a reduction in their rate of photosynthesis. This led to a decrease in turgor pressure in the cells because there was less manufactured sugar. In turn, an increase in the relative vacuole volume and an increased buoyancy resulted, bringing them back up into the region of maximum photosynthesis (Figure 2.6). The vertical movements in both directions were enhanced by turbu-

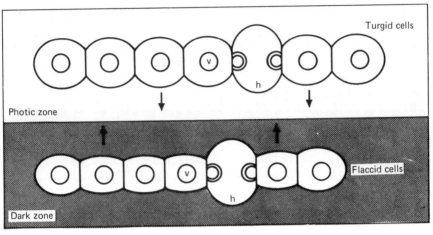

Figure 2.6 Regulatory movements of *Anabaena circinalis* in relation to differing cell turgidities and buoyancies. Highly diagrammatic — for simplicity the numerous gas vacuoles in each cell are represented by one (v) only. Walls of the heterocyst cells (h) do not alter (after data of Reynolds, 1972)

lence but once this was reduced by the sudden onset of calm, fine weather, the generally buoyant nature of the population brought it to the surface, where it became lodged in a film under strong illumination which was unsuitable for satisfactory photosynthesis. Under these conditions turgor pressure of the cells decreased and the relative vacuole volume increased, thus strengthening the buoyancy of the whole *Anabaena* film. Light breezes later deposited the film at the edge of the mere.

Nitrogen Fixation by Blue-green Algae

Algae which can utilize the nitrogen dissolved in water as a starting material for their protein synthesis are known at present only from the blue-green group, the Cyanophyceae (Myxophyceae). Although the contribution which these nitrogen-fixing forms make to the overall income of nitrogen into a water body is generally believed to be small, there are occasions in the annual chemical cycle in lakes when the capacity to carry out the reaction may be of crucial significance and enable the alga to survive and grow. Such occasions occur when the concentration of ammonia and nitrate (combined nitrogen) in the water is low. In certain special enviroments, for example, in paddy fields used for agriculture in Asia, nitrogen fixation by blue-green algae can make such a contribution to the fertility of the system that inoculation is practised deliberately.

For almost a century it had been known that many filamentous blue-green algae such as *Anabaena* had a small proportion of cells which were different in their size and structure to all the others. These were termed heterocysts, and speculation about their function embraced the possibilities that they represented a special kind of reproductive cell or even a point of weakness where the filament could break easily and thus divide vegetatively. Evidence accumulated over the last few years suggests that the heterocysts are in fact the sites of nitrogen fixation while the general cells of the filament have no fixation capacity. The main points of the evidence are now discussed below.

All blue-green algae known to fix nitrogen possess heterocysts, and algae with no heterocysts have no such activity. Addition of ammonia, which is known to inhibit the synthesis of the nitrogenase system involved in fixation, completely inhibits the formation of heterocysts. Other sources of combined nitrogen partially inhibit nitrogen fixation and partially inhibit heterocyst formation. If an alga grown in an ammonium medium is transferred to a medium free from combined nitrogen, nitrogen fixing activity and heterocysts develop in a parallel manner. Using both cell-free extracts and the intact alga, oxygen has been found to inactivate the nitrogen fixing system and it therefore appears important that the latter system should be spatially separated in the alga from the photosynthetic regions, which produce oxygen. Heterocysts, although they contain chlorophyll-a, have no phycocyanin and are thought to be incapable of fixing carbon dioxide and producing oxygen. For final proof of the whole thesis it is desirable that nitrogen fixation should be demonstrated, by the uptake of ^{15}N or by the more recent acetylene reduction technique (see below), in pure preparations of isolated heterocysts. Such preparations can be obtained by a differential cell smashing technique but attempts to demonstrate nitrogen fixation with them have not yielded consistent results. Evidently the technique damages the heterocysts in some way, or only old, inactive heterocysts survive it.

Two main methods have been used to assess the amount of nitrogen fixation brought about by planktonic algae as these grow in lakes. In the radio-isotope method the water samples, which include the algae, are flushed with a carbon dioxide/oxygen/argon mixture to remove naturally dissolved nitrogen. This is then replaced with $^{15}N_2$ and the samples, one litre or more in volume, are then lowered back into the lake to be exposed to light at the same depth at which they were originally collected. After twenty-four hours incubation in the lake, the algae are filtered out of the samples and the $^{15}N_2$ enrichment into the cells is estimated by using a mass spectrometer. The second method for the assessment of nitrogen fixation is based on the reduction of acetylene to ethylene which is catalysed by the nitrogen-fixing enzyme complex (nitrogenase). After flushing as described above, a gas mixture containing acetylene is introduced into the algal sample and subsequently analysed for ethylene by gas chromatography.

In a study at Clear Lake, California, nitrogen fixation was considered in relation to the occurrence of blue-green algae and their heterocysts, and also to the chemical condition of the water.[9] The sampling was concentrated, thirty two stations being visited within two hours on any one day, and this made it possible to relate very local variations in algal and chemical distributions to the fixation pattern. Thus on one particular day, in August, nitrogen fixation was high in one particular bay of the lake but almost negligible elsewhere. This pattern was directly related to the amount of *Anabaena* present and to the number of heterocysts which this alga contained. The blue-green alga *Aphanizomenon* (Figure 2.2 **P**) was also present in the lake at the same time but its distribution showed no direct relationship to the fixation pattern, and it was noted that *Aphanizomenon* had no heterocysts at this time. At other times when *Aphanizomenon* was in a different physiological condition and did have heterocysts it too could make a contribution to nitrogen fixation. Also, an inverse relationship was established between nitrogen fixation and the presence of nitrate nitrogen in the water. When two large sections of the lake in September were compared, nitrogen fixation was observed to be high with a NO_3-N level of 21µg/l in one section, but it was low with a NO_3-N level of 162µg/l, in the other. This inverse relationship between fixation and dissolved nitrate is also apparent in Windermere where estimations made throughout an annual cycle showed that peaks in fixation coincided with troughs in nitrate concentrations; there was little or no fixation when NO_3-N stood at more than 300µg/l (Figure 2.7).[10] It is apparent that the algae satisfy their nitrogen requirements from orthodox sources but turn to fixation as a convenience when the latter diminish. Although the presence of nitrate is deleterious to fixation, the presence of dissolved nitrogen in organic form is not, in fact it even seems to be desirable to act as a 'starter' in the reaction.

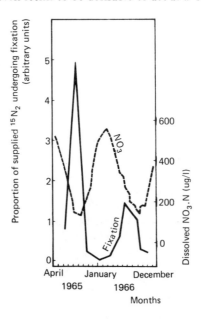

Figure 2.7 The relationship between dissolved nitrate and the extent of nitrogen fixation by algae in Esthwaite Water (after Horne and Fogg, 1970)

The general absence of appreciable fixation in unproductive, oligotrophic lakes has been attributed to their dearth of dissolved organic nitrogen; a general shortage of phosphate is also a contributive factor. In richer, eutrophic lakes, where fixation can be important, a low supply of nitrate together with a sufficient supply of organic nitrogen — the right background to fixation — is often the regime existing at the end of the spring algal bloom and in the late summer and autumn. These are the times when fixation is most likely to occur, but exceptionally it may even take place in the winter, when nitrate is initially at the high level usually recorded during this season. This is when the demand of a large algal crop builds up a requirement which cannot otherwise be satisfied. Such a large algal crop may be of extremely local occurrence in the winter. In October and November, dense blooms of blue-green algae can form overnight in Mitchell Wyke, a small bay of Windermere only 100 m across. No fixation occurs there during the hours of darkness because the mechanism is dependent on photosynthetic phosphorylation. During the following day intense fixation occurs but the reaction soon slows down as the algae disperse into the main body of the lake.

The recent investigations at Clear Lake, California, have suggested that nitrogen fixation by blue-green algae contributes 43 per cent of the nitrogen available for growth in this water body. This is a much higher proportion of the nitrogen budget than that demonstrated in previous studies elsewhere. The large contribution from algal fixation was suspected for this lake before it was investigated in detail, because the nitrogen in the outflow stream greatly exceeds that estimated as being contributed by the inflow sources.

Algal Toxins

Some algal species have the capacity to produce and eventually release toxic metabolites. These can accumulate in the environment to such an extent that they cause the death of fish and other aquatic animals. The toxins produced by *Gymnodinium* and by blue-green algae such as *Microcystis aeruginosa* are endotoxins and are only released after distintegration of the parent cell. *Microcystis aeruginosa* sometimes grows prolifically in British waters; for example, there have been reports of this from the Metropolitan Water Board reservoirs at Staines in recent years, causing some concern. However, although the *Microcystis* endotoxin, which is possibly a cyclic polypeptide, is released when the algal cells are disrupted by freezing and thawing and is then lethal when injected into small experimental mammals in concentrated doses, injection of the water in which the alga has grown is quite harmless. Observation has shown that *Microcystis* algal blooms never give rise to mass mortalities of fish or water birds, suggesting that the mild ingestion which must then occur is not lethal. Treatment of reservoir water to remove the alga before it is passed for human consumption obviates any possible danger. Theoretically, there are conditions when a large release of endotoxin from a blue-green alga could occur in natural waters or in reservoirs. There have been accounts of particular parasitic bacteria or viruses attacking algal blooms, causing epidemics and high mortalities in these and disintegrating the algal cells; under these circumstances the toxin might find its way into the water more readily. However, this represents a combination of circumstances for which there is no authenticated record so far.

Exotoxins are also released by certain algae during their life, and the flagellate *Prymnesium parvum* (Figure 2.2 **R**), a member of the Chrysomonadinae, exemplifies this type. *Prymnesium* is essentially a marine alga, but it flourishes in brackish water in Europe and in Israel. In Israel it has become established in all the fish ponds with brackish water and has caused extensive fish mortalities. Because of its economic importance, it has been studied extensively.[11] The lethal effect of the *Prymnesium* exotoxin is exerted on all gill-breathing animals, including both molluscs and fish. The main symptom which follows the immersion of fish in the toxin is a loss of selective permeability in the gills; toxicants which are not lethal to normal fish can then become so. Under these circumstances the fish cannot resist concentrations of copper and other ions in the environment which are normally sub-lethal. In addition to this effect (called 'ichthyotoxic'), the toxin has a direct effect on cells, causing disintegration of both blood corpuscles of the blood system (the haemolytic effect) and of nucleated cells in general. Chromatographic analysis of the *Prymnesium* toxin indicates that it consists of six components, each showing a different ratio for its relative ichthyotoxic and haemolytic activity. The full chemical nature of the components is not known at present but there are indications that proteolipids are present, containing protein, polar lipids and phosphate. Light is essential for the production of the toxin in the *Prymnesium* cells but the toxin itself is inactivated by exposure to light or to ultra-violet radiation. A unique property of the toxin is the effect of co-factors in enhancing or diminishing its activity as a lethal agent. Thus the addition of polyvalent materials such as streptomycin, spermine or other polyamines enhances its activity to the greatest extent, while addition of divalent cations such as calcium has only a moderate effect. The addition of monovalent cations such as sodium has the least effect. When more than one co-factor is present the resultant toxicity is a function of these combined. Thus if calcium is added to a system which includes a high activity co-factor, the overall degree of toxicity moves to a lower level. It is thought that toxicity is highest in conditions of the lowest salinity in nature, because in such situations sodium does not completely replace the more potent co-factors which may be present, as it does when the salinity is high. Although *Prymnesium* is potentially most dangerous in conditions of low salinity, it is itself very vulnerable there because the osmotic pressure of the environment is close to the lower limit at which the cell, essentially a halophytic one, can survive. It has been found that the delicate osmotic equilibrium of the cell can be disturbed by the addition of low concentrations of ammonia, resulting in death by swelling and lysis. The phenomenon is the basis of control measures which have been applied successfully.

Algal Viruses

The discovery that there are viruses which attack and kill algae is very recent, dating from 1963. So far only blue-green algae are known to be susceptible, the first reports relating to a virus infecting strains of *Lyngbya*, *Phormidium* and *Plectonema* (Figure 2.2). This virus is polyvalent in that it can be transferred from one of these hosts to another, and it has therefore become known as LPP, or more generally as Cyanophage. The pattern of infection which has generally been reported for lower organisms, such as actinomycetes or bacteria, is that many of their viruses are specific to small taxonomic groups, even particular species or individual strains. To demonstrate the presence of the algal viruses in the natural environment, water samples are first incubated with a culture of the prospective host alga. If the appropriate virus is present in the water sample, it increases tremendously as one or more host cells are killed and successive releases of the infective agent then occur to attack further host cells (Figure 2.8). The enrichment solution obtained is tested for activity against a pure culture of the particular alga under investigation. This test is made by sowing dense 'lawns' of the algal cells on nutrient agar and making an overlay with some of the enrichment solution, which is first filtered through a membrane to remove all bacteria. If no virus is present, the algal cells sown in the lawns all develop and a complete dense coverage of alga results. If virus is present in the enrichment solution, derived from incubation of the water sample, then cleared areas or plaques develop on the agar as the virus, starting from a single point, expands its area of influence and kills its host, leaving only cell debris in its path. Observations on plaque manifestations of virus can be made with very simple apparatus, although the use of an electron microscope is finally necessary to confirm that these records made with the naked eye do in fact relate to virus activity. In certain water samples which have been investigated, the amount of algal virus present was so small that concentration was necessary before enrichment could be used to demonstrate its presence. Using this system, the water sample was filtered through paper to remove large organisms and then concentrated by dialysis before being tested. In other water samples

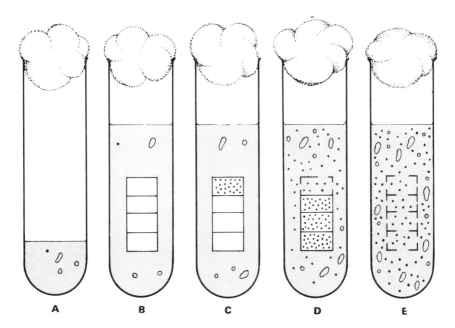

Figure 2.8 Diagrammatic representation of enrichment procedure to recover algal virus from a water sample. A water sample contains a single virus particle and bacteria; B culture of prospective algal host added; C virus reduplicates in an algal cell; D virus released and other host cells attacked; E mass release of virus. Note that bacteria also increase during incubation

13 000 or more virus particles per litre were shown to be present. Workers in Scotland have developed an ingenious method to demonstrate the presence of virus when its occurrence in nature is particularly sparse. The alga is placed as an impregnation on filter paper and then the paper is exposed on small rafts in the selected water body.[12] During exposure the receptive alga picks up the virus, if it is present, from a very large volume of water, and increases it; on return to the laboratory, these samples are enriched and testing for plaque formation is then carried out as described above.

The observations on algal viruses have stimulated much interest. The possibility is opened that algal blooms causing nuisance in reservoirs and fish ponds might be controlled by biological means. However, resistant strains of the host are undoubtedly always present and, although spectacular kills have been reported by Russian workers, the natural situation must represent a more or less balanced equilibrium between alga and virus, with the constant evolution of new sub-strains of both. Another factor which delays the immediate prospect of biological control application is that the most notoriously troublesome blue-green algae, *Anabaena* and *Microcystis*, are apparently not susceptible to LPP and it is problematical at the moment whether viruses do exist which attack them. However, very recent reports suggest that this might be the case.

Aquatic Vascular Plants

Vascular plant species which are partially or totally submerged in water all the year round, and are rooted in or affixed to the substratum, have the life form of hydrophytes. These vascular plants include phanerogams (e.g. *Callitriche, Eleocharis, Elodea, Littorella, Lobelia, Nuphar* and *Potamogeton*), vascular cryptogams (e.g. *Isoetes*) and bryophytes (e.g. *Fontinalis*) (Figure 2.9). The hydrophyte conception is usefully widened to include the Charales (e.g. *Chara* and *Nitella*) of the algae which simulate these higher plants in their growth habit in water. Hydrophytes have internal and external structural features of adaptive significance, for example, a plentiful supply of aerenchyma tissue which allows for adequate internal aeration — a response to low oxygen levels in the water in comparison with the atmospheric situation. Features to allow for both maximum photosynthesis in diffuse light and a maximum rate of respiration at low oxygen levels are: (a) absence of cuticle from stem and leaves; (b) the highest concentration of chloroplasts in the epidermal layer, and (c) leaves much dissected or only a few cells thick. It has been a point of controversy whether total hydrophytes obtain most of their nutrients from the surrounding water or from their roots, which may be very small. A decision on this question would indicate whether the nature of the substratum or some other factor, for example, the amount of available light, is the primary controlling factor in their distribution. The water extraction theory has been favoured by plant physiologists who have used the shoots and leaves of plants such as *Elodea* for ion uptake experiments and the small amount of vascular conducting tissue in the stems, as compared with the sizes of the cortical regions, seems to give support to this theory. However, experiments with *Achyranthes* showed that there is not necessarily any connection

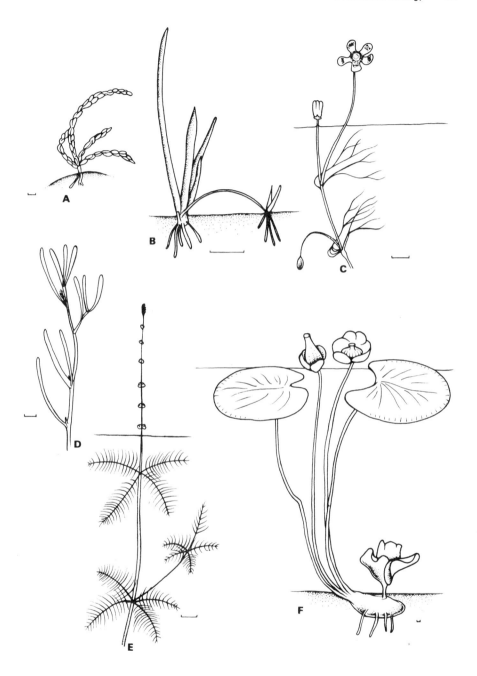

Figure 2.9 Aquatic vascular plants A *Fontinalis antipyretica* B *Littorella uniflora,* showing vegetative reproduction by stolon C *Ranunculus fluitans* D *Potamogeton obtusifolius* E *Myriophyllum spicatum* F *Nuphar lutea* (The scale line represents 1 cm in each case)

between the presence of a vascular system and the route of salt entry. In these experiments the plants were rooted in soil in either a water-deficient or a water-saturated atmosphere. Under the latter regime there was an increase in the size of the cortex and pith and a reduction in the vascular system, although the plants still obtained their salts via their roots. Thus the possibility of total hydrophytes obtaining salts from their rooted portions must be left open. Current thought is that anions can enter readily via the shoots or leaves but may be supplied by the substratum when there is a deficiency in the water at the height of the growing season. It is possible that cations, being relatively insoluble, may enter through the roots, so that a silt-derived mud, rich in colloids, with some degree of metal ion saturation would be expected to be a more advantageous substratum than a sand.

In a recent study of the vegetation of the Scottish lochs consideration was given to this question of the nature of the submerged substrata (termed 'underwater soils') in relation to colonization.[13] In parts of Loch Lomond, and other lochs, large areas of sandy soils occurred and these were largely devoid of any vegetation. However, the *Littorella-Lobelia* community can become established in shallow water on sandy or gravelly soils very low in organic matter, and on such soils a low plant cover is characteristic. The east shore of Loch Maberry (Figure 2.10) exemplified this and *Isoetes* grew in deeper water where the soil was more muddy. On the west shore of Loch Maberry (Figure 2.10), the long unbroken underwater slope, with both sand and mud in the soils, allowed an extensive pattern of colonization with both emergent (*Eleocharis*) and floating-leaved (*Potamogeton natans*) plants represented. In deeper water *Isoetes* occurred again and *Nitella opaca* was present at the greatest depths found to be colonized in this loch. Here, as in lochs and lakes examined elsewhere, *N. opaca* favoured a situation with a low light intensity and soft mud as the soil. Loch Corby illustrated another type of lake margin profile, this time with a well-marked reedswamp on the shore line, dominated by *Phragmites communis* (Figure 2.10). This plant cannot develop on stony or gravelly soils. The enriched status of Loch Corby continued into the water where water lily (*Nuphar lutea*) and a *Potamogeton-Callitriche* community grew from the soft black soil in shallow water.

For certain hydrophytes, e.g. *Fontinalis*, it is difficult to equate colonization with a substratum preference; it may grow either on rocks in shallow water or on soft muds in deeper water. Here we have to consider the possibility that some other factor, for example the abundance of free carbon dioxide in both these situations, may determine the distribution. On deep water soils of the gytta type, characterized by a grey to black colour and good decomposition of organic matter of autochthonous derivation (decomposing algae and aquatic plants) *Fontinalis* competes with *Elodea* and these two plants seem to have a mutually exclusive distribution.

From the foregoing examples and discussions it will be apparent that the nature of the substratum is generally of great importance as a determinant of the type of colonization shown by hydrophytes in lakes, a deduction first made by Pearsall from studies in the English Lake District.[14] The idea that the various communities of hydrophyte and reedswamp vegetation recognized may represent stages in a natural (autogenic) succession, such as that which occurs when woodland is cleared and

Freshwater Biology 41

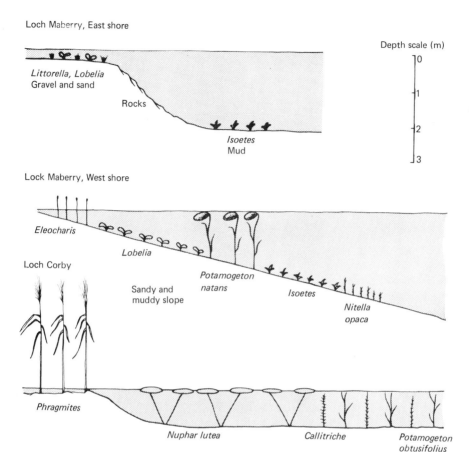

Figure 2.10 The distribution of aquatic vascular plants in two Scottish Lochs (adapted from Spence, 1964)

recolonized, has been under consideration for a long period. In the recent Scottish study the opportunity was taken to examine and photograph the exact sites on the shore margins of a number of Scottish lochs which had been examined and photographed by the botanist West fifty years previously, the objective being to pin-point any obvious changes. These pairs of photographs illustrate very little change or no change at all in a number of cases and it would appear that these represent an edaphic climax of the vegetation in that particular situation. Thus the evidence of autogenic succession is slight. However, in situations where silt has accumulated from the depositions made by rivers there was spectacular evidence of considerable change. Thus at the point of inflow of the River Urquhart into Loch Ness a delta has built up during the fifty years, and it is clear that plant succession from hydrophyte to reedswamp and from reedswamp to woodland has occurred. A similar succession in the same time span has occurred at the inflow of Black Beck into Esthwaite Water in the English Lake District.

Aquatic Vascular Plants as Nuisance Organisms

Various categories of the vascular plants have representatives which have become nuisance organisms in fresh water, but the completely submerged forms are generally the least troublesome in this respect. The most severe problems are caused by the free-floating species. However, even a reduced form such as *Lemna* can be troublesome, as it is when growing amongst irrigated crops in Ceylon, India, Malaysia and the USA. The free-floating species can form vast mats which block drainage channels and hydro-electric installations and can hinder navigation and fishing on both lakes and rivers. The most spectacular examples are plants which have invaded new areas; where they have found favourable climatic and environmental conditions and an absence of natural competition.[15] Under these conditions an explosive expansion of growth can occur. *Eichhornia crassipes*, the water hyacinth, is a native of tropical South America, especially Brazil and Venezuela (Figure 2.11). During the last eighty years, due to accidental and even deliberate introduction it has become a problem plant in Africa, the far East and the southern states of the USA. Its introduction into the latter areas apparently stems from an International Cotton Exposition held in New Orleans in 1884. Here members of the delegation for the Japanese Government distributed water hyacinth plants of South American derivation as souvenirs. Prized for their beautiful exotic blooms, the plants were cultivated in garden and farm ponds, first locally, and then more widely in Louisiana. Escape into the lakes and watercourses of the Mississippi delta was rapid and extensive growth soon occurred. The effect on water transport, vital to the economy of this area for the movement of produce, has been disastrous and so far unalleviated. In Africa there are fears that the water hyacinth, if uncontrolled, might completely block the navigation channels of the White Nile. Here herbicides have been applied to control the weed with some success.

Salvinia auriculata (Figure 2.11), a free-floating fern, also originated in South America. Introduction into Africa has lead to a particularly worrying situation at the new Kariba Dam on the borders of Rhodesia and Zambia. The plant was first noticed in the River Zambesi in 1949, upstream of the dam. When the newly created lake began to fill, in 1958, it was quickly seen that an ideal stable sink had been made for the plant to proliferate in. En route to Lake Kariba from its upstream growth centre, the plant evidently survives passage over the Victoria Falls, where it drops in a colossal mass of water. Most plants are seen to be killed by this impaction but apparently fragments survive and pass on to the lake. It was hoped initially that the *Salvinia* invasion might be a temporary event only, induced by the initial high concentration of nutrients, which is always a feature of water impounded to cover land which was previously productive. However, the established lake receives drainage from fertile land and this seems to have maintained the level of nutrients to such an extent that the plant has continued to grow. A survey in 1966 indicated that 10 per cent of the lake surface was covered by *Salvinia*, and in certain areas secondary

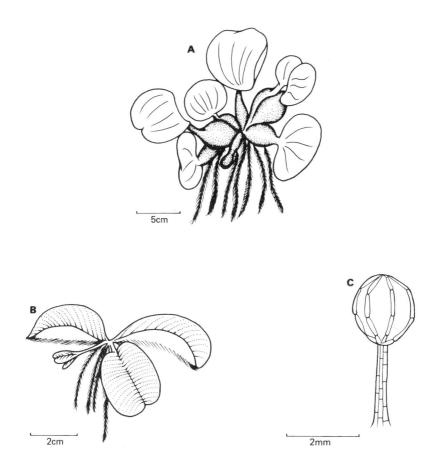

Figure 2.11 **A** *Eichhornia crassipes,* the water hyacinth. Young plant with reclining leaves and petioles swollen into buoyant floats. **B** *Salvinia auriculata,* a water fern, showing water-repellent hairs on the upper surfaces of the leaves; **C** detail of one of the hairs (after Sculthorpe, 1967)

invasion of the plant mats by sedges and other plants had occurred, to form stable platforms of growth (sudd) capable of supporting the weight of a man. There are several deleterious effects of *Salvinia* as it grows in Lake Kariba. There is great loss of water through transpiration and evaporation, particularly if sudd builds up. The commercial inland fishery established on the lake is handicapped directly by the obstruction encountered to boats and nets. More indirectly the plant mats reduce the light and oxygen in the water beneath them, prevent planktonic algae from growing, and disturb the food webs which culminate in fish production.

In Britain the clearance of the excessive growth of aquatic vascular plants in waterways is a matter which concerns the River Authorities; more than two million

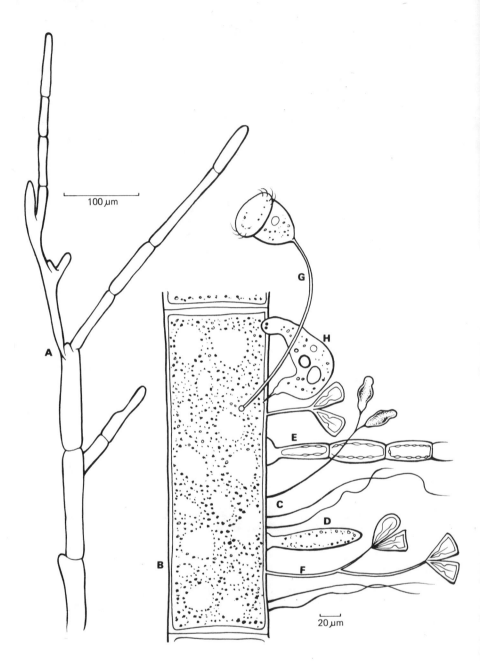

Figure 2.12 *Cladophora* **A** plant outline with growing points **B** single cell from an older section showing loose-network chloroplast and the attachment of numerous epiphytes on the surface: **C** filamentous bacteria **D, E** algal germlings **F** stalked diatoms **G** stalked ciliate **H** rotifer

pounds is spent annually on this. The alga *Cladophora* (Figure 2.12) is often included in these clearance activities since it may grow to massive proportions and interfere with the enjoyment of recreation, especially fishing.

References

1. D.S.I.R. 1960. Water pollution research, 1959. HMSO, London.
2. G. E. FOGG, 1971. *Arch. Hydrobiol. Beih. Ergebn. Limnol.* **5**, 1-25.
3. J. W. G. LUND, 1950. *J. Ecol.* **38**, 1-35.
4. M.J. BURGIS, J.P.E.C. DARLINGTON, I.G. DUNN, G.G. GANF, J. J. GWAHABA and L. M. McGOWAN, 1973. *Proc. Roy. Soc. Lond. B.* **184**, 271-98.
5. J. KALFF, H. E. WELCH and S. K. HOLMGREN, 1972. *Verh. Internat. Verein. Limnol.* **18**, 250-56.
6. J. VERDUIN, 1972. *Verh. Internat. Verein. Limnol.* **18**, 105-112.
7. F.E. ROUND, 1964, *Algae and Man*, Plenum Press, (New York), pp. 138-84.
8. C. S. REYNOLDS, 1972. *Freshwat. Biol.* **2**, 87-106.
9. A. J. HORNE, J. E. DILLARD, D. K. FUJITA and C. R. GOLDMAN, 1972. *Limnol. Oceanogr.* **17**, 693-703.
10. A. J. HORNE and G. E. FOGG, 1970. *Proc. Roy. Soc. Lond. B.* **175**, 351-66.
11. M. SHILO, 1967. *Bact. Rev.* **31**, 180-93.
12. M.J. DAFT, J. BEGG and W.D.P. STEWART, 1970. *New Phytol.* **69**, 1029-38.
13. D.H.N. SPENCE, 1964. *The Vegetation of Scotland*, Oliver & Boyd, pp. 306-425.
14. W.H. PEARSALL, 1920. *J. Ecol.* **8**, 163-201.
15. C.D. SCULTHORPE, 1967. *The Biology of Aquatic Vascular Plants*, Edward Arnold.

Chapter 3
Invertebrate Animals

Invertebrate animals in fresh water exhibit an enormous range of structural and physiological diversity. Their diet ranges from microscopic detritus and bacteria to algae and plant material of all kinds, to food of animal origin and, finally, in the species recognized as carnivorous, to whole living animals. The carnivorous habit may be so extreme as to include cannibalism. The search for suitable food will affect the distribution of the animal and aspects of its life-cycle. A mud-dwelling animal feeding unselectively in this background will be consuming and evacuating food which generally has a low calorific and proteinaceous content. The growth-rate will be slow. On the other hand a carnivorous animal will benefit from the assimilation of food with a greater nutritional value and will grow faster — if it can find its prey!

In this Chapter the diversity of invertebrate life is recognized by considering representatives of three major animal groups, namely the Protozoa, the Crustacea and the Insecta, giving some details of the relevant environmental backgrounds. More invertebrate animals will be considered in Chapter 5, where the special fauna of flowing waters will be examined.

Protozoa

The Protozoa are always present in fresh water and although their single-celled nature might imply uniformity they do in fact exhibit a wide diversity of body structure. Important genera of the freshwater Protozoa are *Amoeba, Actinophrys, Bodo, Paramecium, Loxodes, Vorticella* and *Carchesium* (Figure 3.1). *Amoeba* is familiar as a naked amoeboid form which consumes its food by fairly unselective engulfment, wrapping mobile pseudopodia around it in the process. *Actinophrys* is fundamentally similar but the pseudopodia are long and rigid; they extend from the spherical body and give the whole animal the appearance of a miniature sun with its rays. Food materials such as fungal spores are trapped by the pseudopodia and funnelled down into the body where they remain visible for a while before being digested. *Bodo* is a flagellate with one or two flagella. A more rigid external shape is adopted by species of *Paramecium* and members of the other genera mentioned — these are all ciliates. The cilia line the gullet, into which the food is channelled, and may also beat in unison on the external body surface and move the whole animal. *Paramecium* and *Loxodes* species are free-swimming ciliates while *Vorticella* and

Freshwater Biology 47

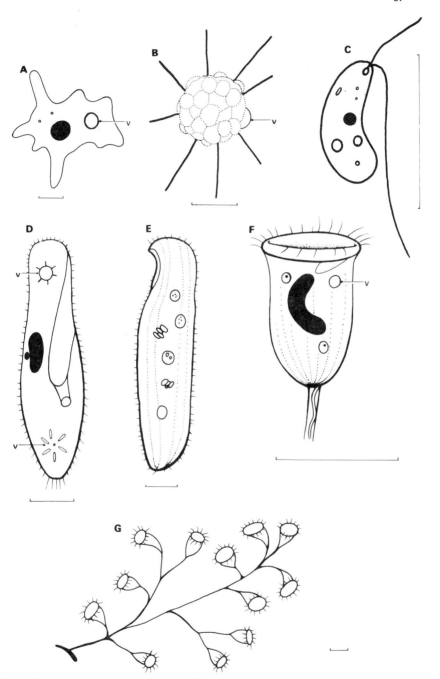

Figure 3.1 Protozoa A *Amoeba proteus* B *Actinophrys sol* C *Bodo caudatus* D *Paramecium caudatum* E *Loxodes magnus,* with *Scenedesmus* cells being ingested (see text) F *Vorticella alba* (after Curds, 1969) G *Carchesium polypinum.* Where they are distinguished, the nuclei are rendered in black, the contractile vacuoles indicated by (v). (The scale line indicates 50μm)

Carchesium species are attached forms, with the latter assuming a colonial habit if the stalks from several individuals unite. The free-swimming forms can be slowed down for microscopic examination by the application of a 1 per cent solution of nickel sulphate, which acts as a slow narcotic.

Vorticella and *Carchesium* become visible to the naked eye when they appear as delicate white felts, simulating fungi, on the surfaces of twigs and branches of trees which have fallen into well-aerated water. Branches standing supported above the mud surface are particularly favoured for this growth. Many of the other freshwater protozoa can be obtained directly from nature or from enrichment concoctions, which build up numbers to such a level that a microscopic examination becomes an easy exercise. The yields from these enrichments often provide evidence of a definite sequence of different protozoans — a succession — as the growth material becomes increasingly degraded.

In one study of protozoan succession, a mixture of animal faeces, tea leaves and yeast was left to develop its own micro-fauna in well-aerated conditions in tanks of water, with constant circulation. The succession obtained was first flagellates, including *Bodo*, then free-swimming ciliates, including *Paramecium*, and finally sessile ciliates, especially species of *Vorticella*.[1] In modern sewage-treatment plants which employ the activated sludge process, the system used for degrading the material is essentially similar to that just described, but on a very large scale and incorporates forced aeration and re-circulation to prevent the build-up of anaerobic pockets. The ciliate fauna which develops is very mixed and includes carnivorous species feeding on other flagellates and ciliates. However, the ciliates which feed on the bacteria have the major role in the degradation process. As the bacterial content of the sludge is reduced, its demand for oxygen (Biochemical Oxygen Demand) falls until the material can be released as an effluent into a river with no catastrophic effects on its flora and fauna.

The activated sludge process runs continuously and, although there is some succession of ciliated forms, the particular species of *Vorticella* which dominates the system at equilibrium is an important indicator of the quality of the effluent being discharged. Thus if *Vorticella microstoma* dominates, the degradation is incomplete and the effluent will have an appreciable oxygen demand; if *V. alba* dominates, the degradation is satisfactory and a good quality effluent is being produced.[2]

In a more natural environment, that of a pond, the growth cycle of one particular ciliate, *Loxodes,* has been investigated in detail. *Loxodes* fed on the algal 'soup' present in this chemically rich situation throughout the year. It was even possible to estimate the digestion rate since the ingested cells, especially those of the green alga *Scenedesmus*, were visible through the body wall.[3] *Loxodes* attained maximum numbers between the months of October to May, and there were then more than 200 individuals per 0·1 ml of the surface sediment. At this time it and all the other ciliates present were confined to the top centimetre of the sediment and it appears that deep penetration into the mud is exceptional. Again, during this winter and spring period, *Loxodes* never moved up into the water to become part of the plankton. Since the pond received farm-yard effluent with a high oxygen demand, the summer situation was that the lower half of the water became anaerobic, although the pond was only 4 m in depth. *Loxodes* cannot survive indefinitely without oxygen, and benthic conditions were clearly unfavourable at this time. In addition to the oxygen depletion, there was a build-up of toxic substances such as ammonia and sulphide. However, although they left the benthic mud itself, it was unexpected that the animals were still found in the lower anaerobic water. It was deduced that they must have made periodic excursions upwards to the aerated epilimnion during this time, and parallel experiments showed that *Loxodes* could endure short periods, from 5 to 10 h, without oxygen. Therefore the following summer picture became apparent. *Loxodes* grazes in the hypolimnion and has some biological advantage there in that it has no competition from animals which lack endurance of oxygen deficient conditions. When autumn and winter return the purely benthic situation is favoured once more (Figure 3.2).

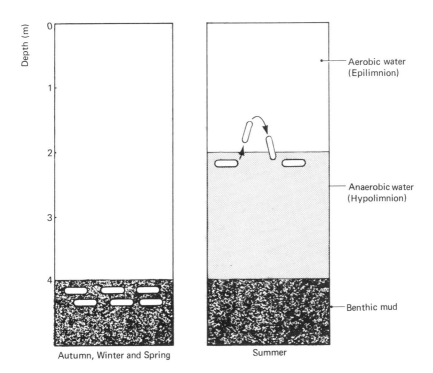

Figure 3.2 The seasonal distribution of *Loxodes* in Priest Pot pond showing the deduced 'aeration excursions' of this ciliate (from data of Goulder, 1972).

Crustacea

The small crustacea of the freshwater zooplankton comprise two major groups, the Copepoda and the Cladocera. Whereas in the Copepoda the eggs are carried by the adult females externally in egg sacs, the situation in the Cladocera, exemplified by

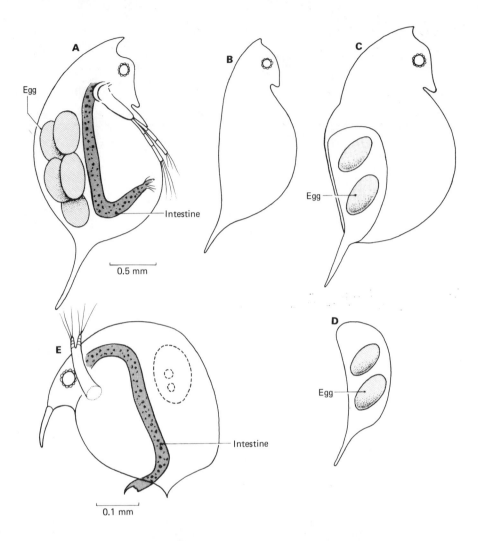

Figure 3.3 Cladocera **A–D** *Daphnia* (after Scourfield and Harding, 1966) **A** parthenogenetic female **B** outline of male showing smaller size **C** outline of female with sexually produced eggs in ephippium **D** shed ephippium **E** *Bosmina longirostris*

Daphnia, is that the eggs lie within the body cavity up to the time of their release (Figure 3.3).[4] Copepods are small, usually 1-3 mm in length, and develop from eggs into nauplii stages, then copepodids, (copepodites) and finally adults through a long series of moults. The nauplius has fewer limbs and a different shape to the adult.

Diaptomus and *Cyclops* are important freshwater copepods, distinguished by the single egg sac of the former and their paired nature in the latter (Plate 2). A comparison of the growth of copepods in two shallow lakes, one in the north temperate zone and the other in the tropics, has brought out a number of interesting points (Figure 3.4).[5] In both of these lakes a single species dominates the zooplankton, *Cyclops strenuus abyssorum* in Loch Leven, Scotland, and *Thermocyclops hyalinus* in Lake George, Uganda. A comparison such as this has to take account of the fact that the amount of the particular copepod which is present at any one time (the standing crop) is a reflection not only of the temperature-regulated growth and reproduction rate but also of the availability of food and the extent to which the animal is preyed on by other animals higher in the food web. In Loch Leven, *C. s. abyssorum* feeds on algae in its nauplii and copepodid stages and tends to be carnivorous when adult, while *T. hyalinus* in Lake George has an algal diet throughout its life. The larva of the midge insect *Chaoborus* (see below) occurs in both lakes and can consume all stages of both of these copepods but is thought to have

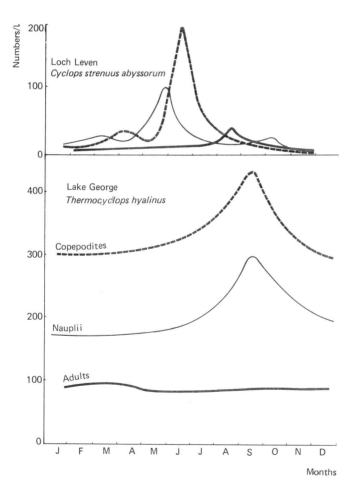

Figure 3.4 Seasonal variation in the numbers of the major divisions in the life histories of *Cyclops strenuus abyssorum* in Loch Leven, Scotland, and of *Thermocyclops hyalinus* in Lake George, Uganda (after Burgis and Walker, 1972)

no decisive effect on the standing crops. In other lakes which have been studied, the presence of *Chaoborus* may greatly affect a standing crop (e.g. Esthwaite Water, discussed below). In both Loch Leven and Lake George, direct predation by fish on the copepods is probably not very marked.

To proceed with the comparison then, the favourable season for copepod growth in Loch Leven, in respect to both food availability and temperature, is the summer. In this season there are 300 individuals of all growth stages per litre as compared with only 4 per litre in the winter. On the other hand, the corresponding Lake George numbers show no marked seasonal fluctuation — they are always high and total 470 to 900 individuals per litre, and here one should note the ready availability of algal food in this lake throughout the year (see Chapter 2). However, since *C. strenuus abyssorum* is slightly longer and heavier than *T. hyalinus,* a more valid comparison of the crops of the two lakes is perhaps made on a dry weight basis. On this basis, once again, Lake George is the more productive. Its standing crop weight, per unit volume of water, is maintained fairly consistently throughout the year, at a figure which is only attained in Loch Leven at the height of the growing season. A factor bearing on the seasonal pattern of abundance in Loch Leven, as compared with the uniformity of numbers throughout the year in Lake George, is a difference in the population structure in the two lakes. In Loch Leven the various instar stages reach maximum numbers in sequence, so at any one time one stage may predominate. For example, the nauplii outnumber the adults just before the latter attain their summer peak. In Lake George, adults are coming to maturity throughout the year and the relative numbers of nauplii, copepodids and adults is always constant. Finally, a comparison of the egg-bearing capacity shows that *C. strenuus abyssorum* in Loch Leven has 38 eggs per female while *T. hyalinus* in Lake George has only 8 eggs per female. Thus in the tropical situation, as compared with the temperate one, the reproductive potential is divided into smaller units and distributed among a greater number of individuals. Similar comparisons, in other groups of freshwater animals, have often yielded the same conclusion.

The introduction of a predator may disturb the balance of copepods in the zooplankton of a lake, and this has happened recently in the Canadian arctic.[6] In Snowflake Lake, *Diaptomus arcticus* was a large dominant species and the local *Cyclops, C. vernalis*, was a small, rare, and essentially fugitive form which could not compete with it successfully and as a result tended to inhabit the lower depths. However, following the introduction of young rainbow trout, *D. arcticus* has been used as food preferentially and this species has been virtually eliminated. *Cyclops vernalis*, benefitting from this removal of competition, has become dominant in the plankton. It remains to be seen whether the introduced fish can now adapt to a smaller food or whether their numbers will decline and eventually come to balance with reviving populations of *D. arcticus*.

Insecta

In biologically rich (eutrophic) lakes in Europe and North America where there is oxygen depletion during the summer season in the deep-water benthos, some invertebrate animals can survive and continue their life-cycles in the muds of this zone. These animals are specialists and they inhabit a world in which there is no light, little or no oxygen for at least one and possibly several months of the year, and a temperature which is always low. In fact it has been said of them: 'It is as if they spend their life in a refrigerator.' Their specialist nature is shown by the fact that only a few species are involved, in comparison with a much larger number of species which can inhabit the benthic mud in littoral situations, where there is oxygen all the year round. In Lake Esrom, Denmark, the important invertebrates in the deep-water (or profundal) benthos are only three in number. One is an insect larva of *Chironomus anthracinus,* and its protracted life in the profundal before maturity can be attained is undoubtedly a reflection of the difficulties of the environment. In this Chapter, in an account based on the classic researches of Jónasson over many years[7], we will consider its life-cycle in some detail. The two other invertebrates which coexist with it are *Ilyodrilus hammoniensis*, a tubificid worm of the Oligochaeta, and *Pisidium casertanum*, a small mollusc. In addition to this trio, the larva of *Chaoborus flavicans*, another insect, also inhabits the mud of the profundal benthos but is not strictly confined there and makes frequent excursions into the water above. All four of these animals occur to some extent in the littoral benthos of the same lake, but in addition representatives of *Polycelis* (Platyhelminthes), leeches (Hirudinea), *Asellus* (Crustacea), *Caenis* (Insecta) and water mites (Acarina) are also present there, giving the overall picture of a very much richer and more varied fauna in comparison with the profundal benthos. Also, in the littoral benthos, the presence of higher plants, which are entirely lacking in the profundal, allows good cover and substrata for a large and varied snail population.

Chironomus anthracinus

The midge insect, *Chironomus anthracinus*, is a common species in European lakes and reservoirs. Although the final phases of the life-cycle occupy only a minute fraction of the whole in terms of time, it is convenient to consider them first.

In Lake Esrom the pupae come up through 20 m of water to the surface at night, in early May. As in other lakes in Europe, in this season of the year conditions are generally calmer at night than during the day and the nocturnal ascent seems to be geared to this; for successful hatching of the adults it is essential that the pupae first float. *Chironomus anthracinus* is prolific in Lake Esrom and in each of three to six consecutive nights — the total span of the emergence — about 500 pupae per square metre are involved. After each nightly ascent of the pupae, the lake surface seems to be covered by a greyish-brown carpet made by millions of wings of the adult insects. After sunrise, as the air temperature rises, the insects fly away. Swarming and pairing begins immediately if conditions are favourable, but once again dead calm is required and sunset is a favoured period of the day for this activity. The fertilized female flies with the hind legs drooping and the brown egg mass is held in an angle near the end of the curved abdomen; when the latter is dipped into the water of the lake, the freed egg mass sinks to the bottom mud. Even at this stage some prediction is possible. If the eggs are deposited into shallow water, the next adult insect generation will probably emerge after one year while if the eggs are deposited into the deepest part of the lake the likelihood is that maturation and emergence will take two years to accomplish.

The egg of *Chironomus anthracinus* is only 0·3 mm long and 0·1 mm wide and the head of the larva to be released from it fits the latter dimension exactly (Figure 3.5). However, the body is coiled inside the egg and it measures 1 mm in length. Because of the rigidity of the chitinous exoskeleton the larva cannot grow continuously to its final length of 16 mm and a series of moults is necessary; at each moult an exoskeleton is discarded. The four successive larval stages (instars) can readily be distinguished, one from another, by the dimensions of the head capsules, and this had made it possible to consider the developmental biology of this insect in great detail. Initial growth of the first instar is rapid and second and third instars can be recovered from the lake within two months. At this time, in June and July, small algae (about 5 μm in diameter) of the Chlorophyceae and Dinophyceae are growing in the epilimnion, and apparently these sediment down to the profundal benthos while still retaining considerable nutritive value, which the animals can exploit. There are of course no benthic algae growing in these deep waters. Since the larval stages are filter feeders, optimum size of the food particle is important, and here we note that these second and third instar larvae cannot engulf the largest diatoms of the lake.

Feeding of the larvae occurs in individual tubes which are lined with a salivary secretion. Despite the difficulty of investigation in the field (and the reluctance of the animal to perform normally in the laboratory), it has been established that different types of tube construction are undertaken in different environmental conditions. Where there is abundant plankton in well-oxygenated water, the tubes are entirely below the mud surface, and this is the situation in the lake littoral. In the lake profundal, however, food is taken directly from the mud surface and the tubes are so orientated that the larvae can extend from it, anchored by their posterior prolegs, and graze over a circular area (Figure 3.5 I). A net of salivary secretion is spread over the mud which is dragged down with its attached particles for final consumption in the seclusion of the tube. In addition to conferring some protection, the tube acts as a channel through which aeration can occur from the entrance, aided by movement of the occupant. This aeration is necessary since, even when the lake water is fully aerobic, the oxygen content of the profundal benthic mud falls to zero only 5 mm below its surface (see Figure 1.3). As the summer draws on, conditions become even more exacting in the profundal benthos. Beginning in late July and extending into August and September (and even exceptionally into October, depending on the vagaries of the particular year), oxygen first falls away steeply and is finally undetectable. During this time *Chironomus anthracinus* shows a diminishing growth rate in its tubes. This is not solely because oxygen lack eventually leads to a general lowering of activity but is also due to a coincidental shortage of suitable algal food. Although the epilimnion of the lake continues to produce planktonic algae, the sedimentation of these to the benthic mud is considerably delayed as compared with the situation in the spring and early summer. This is because the upper portion of the lake is now circulating within itself only and is cut off by a density barrier (at the thermocline) from the lower depths. The upshot is that when

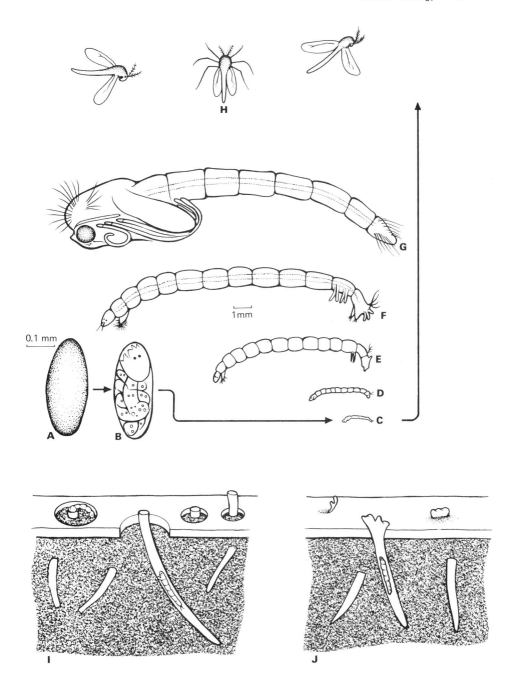

Figure 3.5 The life-cycle of *Chironomus anthracinus* (after Jónasson, 1972) A   fertilized egg
B   developing embryo   C–F   the four larval instars drawn to the same scale   G   pupa   H
midge insect   I   larvae grazing mud surface from constructed tubes   J   winter condition with
tube flaps constricted

the algae do eventually arrive in the vicinity of *C. anthracinus* they are aged, decayed to a considerable extent, and have a low nutritive value. Be that as it may, the situation of the animal in regard to oxygen is of particular interest. Uniquely in the Insecta, *Chironomus* has haemoglobin in its circulatory blood system, undoubtedly correlated with its ability to live in oxygen-poor environments, and becomes so conspicuous to the naked eye of an observer in the later instars that the term 'blood worm' finds ready application. Experiments have shown that the oxygen consumption is constant as the concentration of the gas in the animal's external environment falls from saturation to only a quarter of that value (about 4·0 mg O_2/l). Thus a substantial fall in the external oxygen need have no deleterious effect on the metabolism of the larva. Eventually however oxygen lack does become an inhibitory factor, at an external level of about 0·4 mg O_2/l. This level, or an even lower one (down to zero), occurs in the benthos when anaerobiosis is at its height and during this time *C. anthracinus* is entirely inanimate. The situation is finally alleviated by the autumnal overturn of the lake (see Chapter 1); oxygen returns to the deep benthic regions and growth becomes possible again. There is even an upsurge in temperature, brought about by downward mixing of the warm surface waters, and leading to enhanced activity of the animal. Paradoxically this temperature upsurge in the benthos may be of small extent only if the summer has been warm and calm and much larger if the summer has been cool and windy (see Figure 1.2 on page 12). In the latter case the autumnal overturn occurs early and the benthic fauna may experience many weeks of relatively high temperature before winter begins. Blue-green and diatom algae, the latter constituting the 'autumn maximum', are now growing in the surface waters of Lake Esrom and their ready deposition to the benthos in an undecayed condition aids *C. anthracinus* in its autumnal spurt of growth. These algae are utilized to such an extent that a favourable period of as little as ten days may be sufficient for 80 per cent of the population of larvae to moult to the fourth instar stage. For the first time in its life-cycle the animal is now very vulnerable to predation by fish, which were excluded from the profundal benthos by lack of oxygen during the summer season. The survivorship curve (Figure 3.6) shows that this predation, predominantly made by eels in Lake Esrom, declines during the winter and does not revive in the following spring because the fish then move to littoral waters. *Chironomus anthracinus* itself grows very little in the winter season and during this time the larvae often construct special 'tube flaps' (Figure 3.5 J) which can close the tube entrance and so prevent it from filling with mud if storms occur. The large size of the larvae (now mostly fourth instars) ensures that they can engulf even the largest diatoms such as *Stephanodiscus* from the 'spring maximum' of algal growth in the following April. Growth is very rapid at this time. Body weight is the criterion which determines whether any particular larva will undergo transformation to a pupa and emerge as an insect in May or whether emergence will be delayed for another whole year. On an average basis, larvae which pupate weigh 13 mg while larvae which continue to grow as such weigh 10 mg. In Lake Esrom the fraction of the *C. anthracinus* population which comes to maturity as an insect after only one year in the profundal benthos is rather small (Figure 3.6), and a regulatory factor in this is the difficulty which each new generation has in becoming established on the

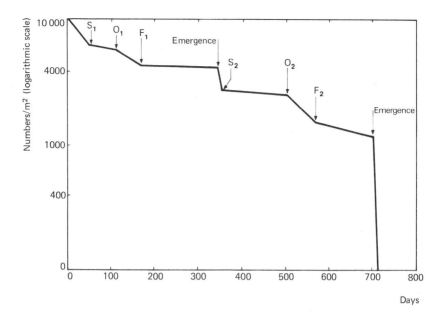

Figure 3.6 Survivorship curve for *Chironomus anthracinus* in Lake Esrom. S_1 and S_2 indicate the beginning of summer anaerobiosis, O_1 and O_2 the advent of autumnal overturn of the lake, F_1 and F_2 the cessation of predation by fish (after Jónasson, 1972)

mud in competition with its own kind. If the fourth instar larvae which remain are sufficiently dense (and they may total 7 000 per square metre) they will consume most of the developing eggs in their vicinity and effectively prevent any large new summer recruitment. Under these conditions the life-cycle is predominantly a two year one. However, if most of the larvae of a population pupate and emerge after one year then cannibalism is not a problem for the new generation and a predominantly one year cycle is established. This occurs in the littoral of Lake Esrom where there is no summer restriction of growth due to anaerobiosis as there is in the profundal benthos.

We can quickly fill in the details concerning the other two invertebrates which accompany *Chironomus anthracinus* in the deep benthos of Lake Esrom and in so doing complete a picture of the whole community which surrounds it. The tubificid worm, *Ilyodrilus hammoniensis*, is very abundant, totalling 25 000 individuals per square metre. In feeding it passes mud unselectively through its intestine, and the low energy content of its food is probably the main reason for its extremely long life-cycle of four to six years. The mud is transported upwards to the mud surface from the deeper layers and hence the animal works in an opposite direction to the *Chironomus* larva. Reproduction occurs when embryos are deposited in cocoons and they develop to maturity in two months or less. Although the activities of *Ilyodrilus hammoniensis* seem unspectacular, there is little doubt that the animal has a large role in the reduction of organic matter made by the whole benthic fauna at the mud surface. *Pisidium casertanum*, the small molluscan representative in the

Lake Esrom profundal benthos, has a respiratory adaptation to profundal living which is unexplained; unlike *Chironomus* and *Ilyodrilus* it does not contain haemoglobin and laboratory experiments show that its oxygen consumption falls markedly when the external concentration of this gas is reduced. It inhabits burrows and like *Ilyodrilus* feeds unselectively in the mud.

The detailed work on *Chironomus anthracinus* in Lake Esrom, placed in context with integrated studies on its fellow community members, makes it possible to draw wider conclusions. For example, from estimates of numbers of animals occurring in the lake and laboratory experiments on respiration of individual species it has been possible to consider the total amount of oxygen consumed by each member of the community. *Chironomus anthracinus* is estimated to consume 75 litres of oxygen per square metre per year as compared with 10 litres for *Chaoborus flavicans* and only 1.5 litres for *Ilyodrilus hammoniensis*. Finally, an estimate of community respiration can be made and this is a component of the total energy flow through the whole lake. Energy flow estimations make it possible to compare lakes on a wider scale and gauge the efficiency of these water bodies as productive units — a vital consideration if a fisheries exploitation is contemplated, for example.

Chironomus anthracinus, a well-established species, has been the subject of basic biological research. Other species of chironomids are much less well-known but may suddenly become a focus of current interest. An example may be cited. From 1971 onwards complaints were received from inhabitants of Southend-on-Sea, Essex, that 'worms' were coming through in the domestic water supply. These 'worms' were identified as the larvae of a chironomid, and in view of the modern water treatment procedure, which included filtration, it was mystifying how these escaped from the reservoir. It was subsequently shown that the particular species involved, a hitherto unknown one, was not escaping from the reservoir but was actually increasing in the mains water pipes. Under these circumstances the final stage in the life-history, the flying insect, was omitted and eggs were produced parthenogenetically within the pupa. They developed into larvae feeding on detritus in the pipes. Further research is in progress and the problem of eradication is not yet satisfactorily resolved since this species has a very short generation time (a few weeks only) and can rapidly replace populations which are almost totally destroyed.[8]

Chaoborus flavicans

Although it may seem repetitious to consider another insect larva which inhabits the profundal benthos of lakes, *Chaoborus flavicans* is so remarkable that its omission would be unthinkable. This larva is commonly known as the phantom or glass larva when it is large (1 cm long) and conspicuous in its fourth instar. It is almost totally transparent and excites wonder whenever it is found. Unlike *Chironomus anthracinus* it is an active swimmer and paired air sacs at each end of its body play a part in governing its hydrostatic relations as it rises and falls in the water column. At night it rises to the lake surface but during the hours of daylight it is found mostly in the benthic mud. In dull weather however it tends to be pelagic over the whole 24 h period. Its respiratory system is remarkable since it can apparently migrate to the lake bottom at all seasons irrespective of the oxygen regime. The pattern of movement shown by this larva is undoubtedly related to its feeding habits since it is carnivorous on small crustaceans in the zooplankton (Figure 3.7).

In Esthwaite Water, in England, it has been noticed that in years when the larva is particularly abundant there are only small amounts of *Cyclops strenuus abyssorum* in the plankton. Conversely, in the poor years for *Chaoborus,* the *Cyclops* is much more numerous. This is highly suggestive of a close predator-prey relationship, and in laboratory experiments *Chaoborus* has consumed 150 *Cyclops* nauplii in 24 h.[9] Although *Chaoborus* is fairly unselective in its feeding, the *Diaptomus* and *Daphnia* which occur in the same lake are less vulnerable and their annual numbers do not show similar inverse relationships, probably because they do not migrate downwards so deeply during the hours of daylight as does *Cyclops*. It is known that the *Chaoborus* larva starves to death without regular food and only small concentrations of zooplankton are available in the winter months. It has therefore been suggested that it probably feeds on tubificid worms in the benthic mud at this time. *Chaoborus flavicans* has a one year life-cycle with the adult insect, again a

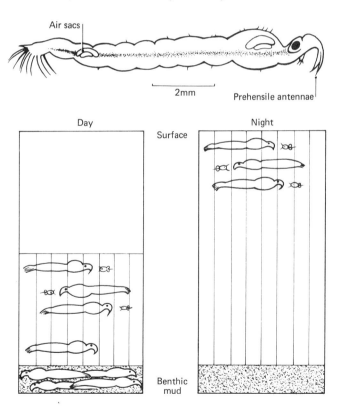

Figure 3.7 *Chaoborus flavicans* larva showing how its diurnal pattern of movement follows that of its prey, *Cyclops*

midge, emerging in July. This rapid turn-over of a sizeable animal reflects the advantage of obtaining a richer food supply than that which is available to the invertebrates more strictly confined to the deep benthos.

References

1. C. R. CURDS, 1964. *Oikos* **15**, 282-9.
2. C. R. CURDS, 1969. *Water Pollution Research Technical Paper 12*, HMSO.
3. R. GOULDER, 1972. *Freshwat. Biol.* **2**, 163-76.
4. D. J. SCOURFIELD and J. P. HARDING, 1966. *Freshwater Biological Association, Scientific Publication 5.*
5. M. J. BURGIS and A. F. WALKER, 1972. *Verh. Internat. Limnol.* **18**, 647-55.
6. R. S. ANDERSON, 1972. *Verh. Internat. Verein. Limnol.* **18**, 264-8.
7. P. M. JONASSON, 1972. *Oikos, Supplementum 14*, 1-148.
8. D. N. WILLIAMS, 1974. *Water Treat. Exam.* **23**, 215-31.
9. W. J. P. SMYLY, 1972. *Verh. Internat. Verein. Limnol.* **18**, 320-6.

Chapter 4
Fish Biology

Fish constitute the ultimate expression of the biological productivity of a freshwater river, pond or lake and economic considerations give them high importance. Salmonid fish, exemplified by the salmon, trout and char are particularly prized as food. An initial distinction of these from the coarse fish is made in their possession of an adipose fin, which the latter lack (Figures 4.1 and 4.2). Important freshwater coarse fish in the British Isles are the eel (*Anguilla anguilla*), the perch (*Perca fluviatitis*) and the pike (*Esox lucius*), and in order to introduce some of the basic conceptions and methods in fish biology a detailed life-history of the pike is considered, based largely on observations made in Windermere.[1]

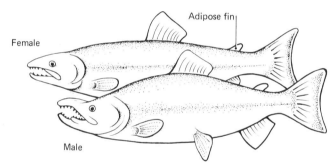

Figure 4.1 Pacific salmon, *Onchorhynchus nerka*

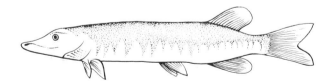

Figure 4.2 Pike, *Esox lucius*

The Pike (*Esox lucius*)

In Windermere, the sexually mature pike arrive at the spawning sites in the month of April, when the water temperature is 6-8 °C. Up to fourteen days or longer is spent in these selected situations and the evidence from tagging and re-capture is that pike 'home' — in the sense of returning repeatedly to spawn at almost exactly the same site in successive years. One specimen was re-captured on its selected site in four consecutive years. Spawning sites occur in sheltered bays, and the eggs are attached to submerged plants, such as *Elodea, Myriophyllum* and *Nitella*. The fertilized eggs, which stick on singly, are about 0·3 cm in diameter and are camouflaged to some extent by their dirty greenish colour matching the supporting plant. The eggs hatch in fourteen to twenty-one days and 1 May has been taken as the biological 'birthday' for pike in Windermere. The newly hatched alevin has a large yolk sac and an anterior sucker which maintains its attachment to the plant (Figure 4.3).[2] Observations on this and the succeeding very early stages have been made in aquaria and show that by four days most alevins are swimming, but rest occasionaly, attached by the sucker. The function of the latter is undoubtedly to keep the alevin from dropping into the mud and silt below, where its chance of survival would be very greatly reduced. At six days old the mouth and anus are open and the sucker is no longer in use, and by nine days old, when it is 1·1-1·2 cm in length, the yolk sac has almost completely disappeared. The young pike is carnivorous, accepting living prey only, and maintains this habit for the rest of its life. The feeding pattern is an excellent example of adaptability, geared to the availability of different prey at different times of the year and to its own increasing size during its life. It prefers food which is about a half to a third of its own length. When its size is up to 1·8 cm, therefore, it selects small Crustacea, particularly the Cladocera *Bosmina* and *Chydorus*, but also the Copepod *Cyclops*. Up to seventeen individuals of *Cyclops* have been counted in the stomach of one pike fry. Insects, especially chironomid larvae and pupae and Ephemeroptera nymphs first enter the diet

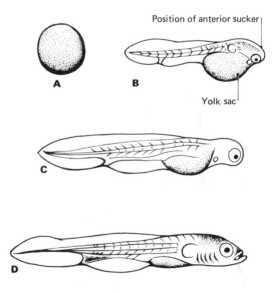

Figure 4.3 Egg and young stages of pike A egg B newly hatched alevin C alevin at six days D alevin at ten days (all x 5) (after Kennedy, 1969)

when the pike is 2·5-3·0 cm in length and are especially important at this stage although they may still be taken much later in life. Fish enter the diet when the pike is about 3·5 cm long, in June, and at this stage body scales are first developed. In June, young perch fry are available from the current year's spawning of this fish in Windermere, and since these occur in shallow water and are very vulnerable they are preyed on extensively. The technique of capture is first to 'mark down' the prey, which is stalked stealthily to within a certain distance. The body is flexed and with the impetus so derived there is a lightning forward dart and strike. The prey is first seized crosswise in the mouth. Here, firmly held by the backwardly projecting teeth, the prey is mouthed so that it faces head first in the gullet of its captor, who then swallows it whole. The capture of prey which would seem to us to be out of the pike's direct line of vision could be due to the great mobility of the eyeball of the pike which allows it to see in almost any direction. The relative sizes of the pike and perch prey recorded during this time were 3·5-9 cm and 1-2 cm, respectively. It must be emphasized that we are considering a Windermere population in this account and that in other localities the feeding pattern may well be different in accordance with differing local conditions. Thus a study of pike in Danish streams shows that these pike turn largely to the 'freshwater shrimp' *Gammarus* at a corresponding stage in the life-history and in so doing become a serious competitor with trout for this food. The evidence for Windermere is that *Gammarus* is only taken to a very limited extent. With respect to this account of Windermere pike, minnows also make a contribution to the food supply in the first summer; they are again available since they have left their winter haunts under stones and are now in shoals in shallow water.

By late summer the pike are already showing clear signs of size selection in their feeding, specimens of up to 5 cm in length taking the smaller minnows of up to 1 cm and specimens in the 12-20 cm size range indulging in the larger minnows of 2-5 cm. As late as October the perch fry in shallow water are still clearly important as food, but after that they are presumably more difficult to capture since they move to deeper water and adopt a less gregarious existence. Our knowledge of the pikes' feeding routine during their first winter is incomplete but the evidence from the examination of stomach contents suggests that perch and minnows continue to be consumed and that small trout first appear in the diet. By the following summer, the larger perch in shallow water, rather than their fry only, can be taken. In the autumn of either this year (age 1) or the following one (age 2) spawning char (*Salvelinus alpinus*) and brown trout (*Salmo trutta*) are attacked and eaten and this predation then becomes a regular annual event in the life of the pike. Char congregate in shallow water in the lake to spawn in November and December, while trout are running through the shallows at the same time en route for their spawning inflow streams. The char are particularly favoured and there is no doubt that they constitute the main diet of adult pike in these two months. To summarize this particular predation picture, examination of stomach contents showed that pike of all sizes from 40-105 cm ate char which were mostly of 27-30 cm, the narrower range of the latter dimensions being accounted for by the fact that these were almost all mature breeding fish.

Considering the predation on trout during the whole life of the pike, the records indicate that there is a strong connexion between the size of the predator and that of the prey.

Table 4.1 *Mean size of trout recovered from the higher length-groups of pike*

Length of pike (cm)	Length of caught trout (cm)
50–59	25·9
60–79	29·4
80–99	31·9
99–105	43·0

Predation on the large trout of Windermere is almost entirely by the larger pike of 60 cm and above. A final note on pike feeding concerns the often quoted generalization that 'a high proportion of empty stomachs is a characteristic of the fish-eating fish'. The Windermere figures indicated that 83 per cent of small pike less than 20 cm were feeding at the time of capture while only 52·7 per cent of the larger pike (20-105 cm) were doing so. Undoubtedly the large size of the foods eaten by the mature fish enable it to endure an interval between each feeding period, while the stomach is empty. Is the pike a cannibal? The evidence shows that it is when other suitable food is not available and it may also become so in crowded aquarium conditions.

Two years after life began as fertilized eggs in Windermere, the pike are sexually mature and by this time there is a difference in mean size between the males and females, around the 40 cm mark, the females being the larger by about 2 or 3 cm (Plate 4). Apparently breeding then occurs in both this and every subsequent year. From a comparison with observations from other countries and localities it is seen that the age of pike at first spawning is dependent on the growth rate. Thus in warm Spanish waters the pike grow particularly quickly and breed after only one year, while in Great Bear Lake, Canada, the fish take five years to attain 40 cm in length and only then are they ready to spawn. A seasonal feeding pattern in relation to spawning occurs in Windermere, where the lowest percentage of feeding fish is found from January to April. The figures for January and February suggest a slowing down of feeding in these cold months, but the lowest of all, in March and April, are probably indicative of a fast associated with spawning time. On the other hand, the high figure for feeding in May probably reflects the hunger and voracity of spent fish building up their body weight. The capacity of pike to withstand internal debilitation at breeding time and to recover from it quickly is evinced by the absence of any sign of heightened vulnerability to disease at this time. As will be discussed later, this is in contrast to salmonid fish which are particularly vulnerable at this time. The great fecundity of pike, that is, the very large number of eggs deposited, has long interested naturalists; for example, Buckland in 1873 recorded that a 28 lb pike had 292 320 eggs and a 32 lb specimen 595 200. The Windermere records suggest that 96 per cent of these eggs by volume are mature while 4 per cent are immature. The latter are small and are destined to be shed a year or more later. A very slight rise in the weight of the ovaries begins in late summer, then from October until the end of December they build up steadily. Increase in their weight continues during January and February, until by the end of the latter month ovaries comprise about 15 per cent of the total body weight. In contrast to this, the testes of the male fish form only 2-4 per cent of the body weight when they are fully built up. When egg counts were made from the female pike of Windermere and related to the weight of the fish it was seen even in the mature fish of any one year class that the variation in this ratio was considerable. However, there was also a suggestion of variation between different years. An attempt was made to relate the observed changes in the number of eggs per gram of fish weight to environmental conditions which occurred one and two years before the eggs were due to be spawned. There was no apparent connexion between temperature and food supply and the number of eggs, but the rise in egg numbers occurred after the population density had been falling for several years. It was suggested that the changes in the egg to body weight ratio could be a population regulating mechanism, dependent on the numbers of the fish present.

The background to the study of mature pike in Windermere, which has involved intensive annual netting of these fish in the winter months, was the realization that there has been a marked increase in the biological productivity in the lake over the last two centuries. There are many signs of this, some of them from the changing nature of the sedimentary deposits, which will be discussed in a later chapter. This is by no means a phenomenon unique to Windermere since many other lakes in various parts of the world are showing the same trend. On Windermere there is evidence both from ancient literature and from local observation which has been handed down through several generations that this change in biological productivity was accompanied by a change in the balance of the fish species. The coarse fish pike and perch became more abundant while the salmonid fish, particularly char and trout, decreased in numbers. In 1944, therefore, it was decided to keep the pike in check in the interest of the more highly prized salmonid fish and in so doing make an extensive investigation of the population dynamics of mature pike. Pike netting is carried out in the shallow bays where experience has shown that captures can be made. The nets have a 5 inch mesh (2½ inches knot to knot) and are 30 yards long and 10 feet deep. This particular mesh size was chosen to allow smaller fish of all species to escape, particularly trout of up to 3 lb in weight, and although larger trout are occasionally caught by accident, the timing and placing of the netting ensures that it is very selective for the larger pike. The net is essentially a gill net, trapping the fish just behind the head, but to some extent it is loose and hangs in folds. Thus it acts as a 'tangle' net and large pike with a body diameter even larger than that of the extended mesh are caught since their struggles to escape often lead to entanglement with the jaws or fins. Therefore the net imposes a lower size limit for capture but not an upper one.

This question of net selection is such an important one in all fishing operations that it will now be considered in more detail. Experience has shown that Windermere pike smaller than 55 cm in length will not be held and only a certain proportion between 55 and 65 cm will be. Above 65 cm almost all the pike coming into contact with the net will be held. This introduces bias into a sampling aimed at elucidating the sex, size and age structure of the total population. When one considers the age of the pike in relation to capture, very few fish of 1-2 years are caught, as could be predicted. This study has shown that for the year-classes 1942-59 — that is, the pike which were born in each of those years — the mean length of females at age 3 years ranged from 55 to 59 cm. Even at the highest figure less than half the fish of this group would be expected to be held by the net, so catches for this age are definitely biased on account of net selection, the bigger fish being caught and the smaller escaping. At age 4, however, the mean lengths of females ranged from 65 to 70 cm in the different year-classes. At the smallest mean length about 80 per cent of the fish would be held by the net and for all other mean lengths the percentage would be higher. The net selection effect on females age 4 years and over is therefore small. Since the males grow more slowly than the females, as mentioned above, the effect of net selection is different for them. At age 3 few males are caught. At age 4 the mean lengths of males in several year classes has varied from 56 to 61 cm, and thus again a considerable proportion would not be caught. A similar but lesser escape occurs even at age 5. Therefore the results for males are considered to be biased up to the age of 6. The effect of this selection is to catch the fastest growing young male fish and thus exaggerate

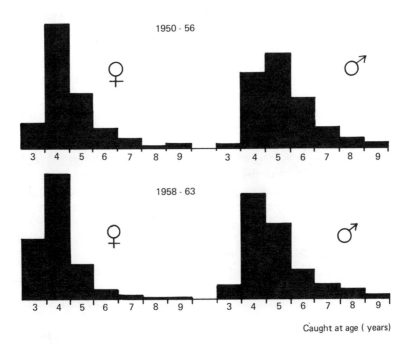

Figure 4.4 Percentage age distribution of pike caught in Windermere in two different periods of years, males and females separately (from Frost and Kipling, 1967)

the growth of the males as a whole in their early years. However, the effect of this selection in the examination of the total population is mitigated by the annual nature of the sampling. Fish not caught in one year will be caught in the next, so eventually the whole of the year-class is sampled. The result of differences between males and females in net selection is shown in the different age distribution when captured, most of the males being 4-7 years old and most of the females 3-5 years (Figure 4.4). How can we tell the age of a pike? How was it possible to produce all the data which has been quoted above on the sizes of the two and three year old fish, bearing in mind the very selective nature of the netting programme? The answers lie in the careful measurement of the fish obtained in the routine netting programme, their ageing by scale and bone measurements, and the back-calculation of growth by mathematical methods.[3] Information and measurement from tagged and released fish subsequently recaptured in the netting programme or caught and returned by local anglers is also extremely valuable. The length of each pike is measured from the tip of the snout to the indentation of the tail. Body scales are removed from midway along the length of the fish, good specimens with clear centres being selected. These are soaked in 4 per cent sodium hydroxide solution overnight and then cleaned with a brush and examined dry between two glass microscope slides. Opercular bones are removed from the outside of the fish, just above the gill cover, using a scalpel to free them. They are placed in hot water and then cleaned with a cloth and stored dry. The information to be derived from opercular bones is clearer if these are also stored dry and left to 'mature' for some months before examination.

An examination of pike scales for ageing purposes shows that a single scale consists of three anterior lobes, which are buried when the fish is alive, and a smaller posterior lobe, which protrudes when the fish is alive (Figure 4.5). Both anterior and posterior lobes have fine rings or circuli. These are not uniformly spaced but are arranged in alternate groups of widely and narrowly spaced circuli, a grouping

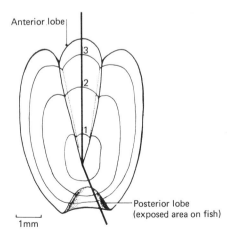

Figure 4.5 Pike body scale, showing annuli along lines of measurement

which is much better defined on the anterior than the posterior part of the scale. The initial supposition is that the production of circuli proceeds at a regular rate throughout the life of the fish, but the production of the narrowly spaced circuli denotes a growth check, termed an annulus, while the production of the widely spaced circuli denotes a period of rapid growth. If these growth checks are of annual occurrence then they may be used to compute the age of the fish from which they were derived. What is the evidence from pike that the scale annuli are of annual occurrence? Generally the evidence is satisfactory. The numbers of annuli recorded on the scales of tagged and returned fish usually agree with the known number of years of absence. However, as a further check, the ages of eighty-five fish of varying lengths were read from both anterior and posterior annuli on the same individual scale mounted for examination. On fish up to five years old the number of annuli was the same both anteriorly and posteriorly but with increasing age there was a discrepancy of up to three years, and in most cases the anterior number was the higher. This pointed to the possibility of 'false annuli' occurring; there has long been the suspicion entertained by various pike investigators that these may be formed. Further investigation confirmed that if ageing was to be made from body scales then the posterior rather than the anterior portion should be read and this invoked the difficulty of measuring the small distance between the annuli on these. Measurement is necessary to make back-calculations of growth in addition to those of age. Accordingly, further growth and age determinations were made from opercular bones (Figure 4.6) and proved much more satisfactory than from scales, and this is the method in current use, although scales still may provide valuable corroborative data.

When plotted arithmetically, measurement of lengths of pike and the lengths of

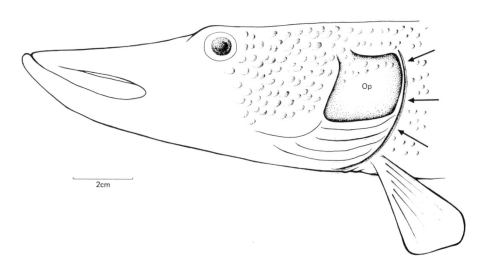

Figure 4.6 Pike, *Esox lucius*, showing position of opercular bone (Op) in relation to head and gill cover (arrows)

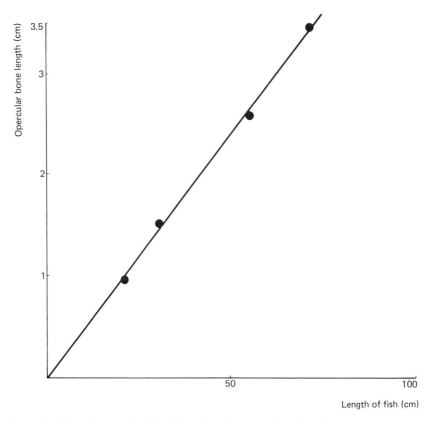

Figure 4.7 Pike: relationship of body length and opercular bone length

their opercular bones give a straight line relationship between the two (Figure 4.7). This implies that the opercular bone grows at the same rate as does the whole fish throughout life, a connexion which does not hold for the anterior body scales, since similar plots give a curve in that case. Fairly clearly distinguished and conveniently distanced for measurement are the opercular bone rings (Plate 3). What is the evidence that they are annually produced, that is, they are true annual rings? Part of the evidence is that opercular bones taken from fish captured in summer and autumn have an opaque white zone at their edge, whereas in winter and spring (until April) there is a transparent dark zone at their edge. This suggests that a seasonal growth pattern occurs. Other evidence from the recapture of fish which have been tagged and at large for a known period of time confirms that this is so and that the opercular bone rings are in fact true annuli. However, there may be difficulty at the thick base of the bone where early annuli can be obscured. This will be discussed below.

In order to consider age and growth in more detail it is convenient to introduce mathematical conceptions at this point. A very useful mathematical conception in studying the growth pattern of pike is that of the Von Bertalanffy equation which describes growth by two parameters, $L\infty$, the ultimate length, and K, the rate (slope) at which the ultimate length is approached. $L\infty$ is affected by the food supply and population density while K is affected by genetical and physiological factors. For pike of certain year-classes (that is groups of pike all born in the same year) in Windermere it was shown that $\delta L\infty = 77$ cm and $K = 0\cdot36$; $\female L\infty = 99$ cm and $K = 0\cdot28$. A graphical transformation of the Von Bertalanffy equation has been made in Ford-Walford plots. These assume that the successive yearly increments added to the length decrease in magnitude in geometric progression, until a limiting value of total length (ultimate length) is approached. In Ford-Walford plots the form of the growth curve is changed by plotting 'length at age n' against 'length at age $n + 1$' and this gives a straight line relationship when the assumptions above are fulfilled. A theoretical example is considered first (Figure 4.8). Only two initial assumptions are made, namely that the length at age 1 is 25 cm and that the ultimate length is 100 cm (such assumptions would roughly match our knowledge of fast growing female pike in Windermere). The 25 cm point on the y-axis is joined to the 100 cm mark on the diagonal which passes through the origin. Length at age 1 is again located, this time on the x-axis, and the age at $n + 1$ (=2) read off as 43 cm on the y-axis. Successive plotted points then give estimations of the lengths of the fish in succeeding years. Since the growth in length of the fish is directly proportional to that of the opercular bone, calculation can convert measurements of the latter, taken in relation to the annuli, into the lengths of the fish at its different ages. Thus it is possible to work back in time and deduce the length of a captured fish when it was younger. This procedure is known as back-calculation. The application of Ford-Walford plots to derive information is now considered.

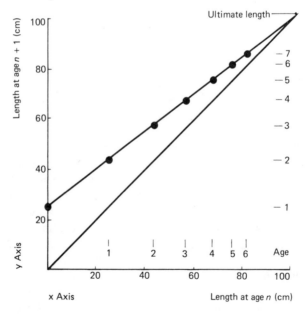

Figure 4.8 Ford-Walford plot of pike growth

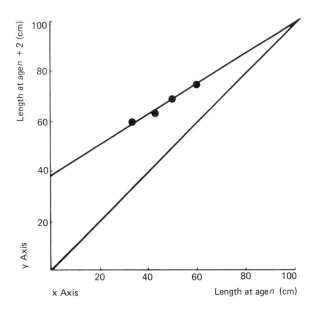

Figure 4.9 Female pike length at age 2

Example 1. Length at age 2 (Figure 4.9)

Four female pike (of different ages) were absent exactly two years between tagging and recapture. The two length measurements (made at time of tagging and at time of recapture) on each fish yielded one plotted point and a line of best fit could then be estimated for the four points. This line cut the y-axis at 38·9 cm, giving this as the estimate of length at age 2, and cut the diagonal through the origin at 98·3 cm, giving this as the ultimate length.

Example 2. Back-calculated growth and missing annulus (Figure 4.10)

The growth of a single female pike was back-calculated from measurements of the opercular bone annuli and this yielded the plotted black points. The length of the fish so back-calculated for the innermost visible annulus was 42·4 cm. A line of best fit was drawn through the black points and this cut the y-axis at 24 cm, giving this as the estimated length at age 1. Length at age 2 was therefore 42·4 cm; the innermost visible annulus was the second annulus, and the fish was age 8. Few pike under the age of 4 have missing annuli, while over 7 almost all have at least one missing with two as the maximum found so far.

Example 3. Growth of male pike (Figure 4.11)

This plot shows the mean lengths at ages 1 to 5 of 54 male pike caught when aged 5 or over. These mean lengths were back-calculated from the opercular bones, using estimates for missing early annuli when necessary. The line of best fit for the plotted points cuts the y-axis at 23 cm, which was the estimated mean length at age

72 Fish Biology

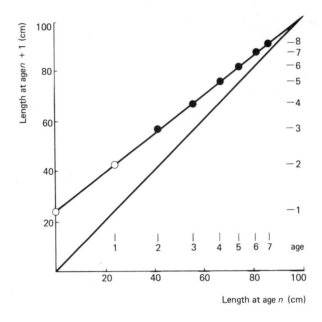

Figure 4.10 Pike: back-calculated growth and missing annulus

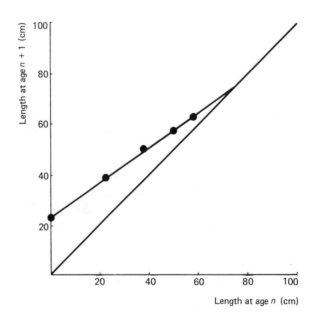

Figure 4.11 Growth of male pike (Figures 4.7 to 4.11 from Frost and Kipling, 1959)

1, and the diagonal through the origin at 75 cm, which was the estimated ultimate length for this group of fish.

The great variation between the growth of individual pike growing in the same environment has frequently been commented on. In the Rybinsk reservoir in Russia, an investigator recorded that the smallest 5 year old specimen weighed 500g and the largest 2045g. A similar variation occurs in the Windermere pike, complicated by a sex difference as mentioned above. Ford-Walford plots give 20 and 21 cm, respectively, as the mean lengths for males and females at age 1, but the data from observations suggests that the range in length of individual fish is 11-35 cm. At age 2 plotting gives a mean length of about 40 cm for the females (see Figure 4.9) and a range of 25-50 cm. The length variations in these and subsequent years have been converted to weights using appropriate calculated weight-length regression equations and it is seen that individual females may differ in weight by about

Table 4.2 *Weight estimations of Pike in their first four years. 0·95 confidence limits of weight in g. (From Frost and Kipling, 1967).*

	Age			
	1	2	3	4
Females	30-180	300-1200	1100-2700	2200-4300
Males	30-180	270-870	850-1700	1500-2300

six times at age 1, four times at age 2, 2½ times at age 3 and twice at age 4 (Table 4.2). It was suspected that some of the variation between the growth of individuals could be ascribed to different climatic conditions, especially summer temperature, which the fish had experienced during their life. This possibility was examined by collecting data from different year-classes. In any one year-class all the fish were born in the same year, although they could have been captured at different final ages. The mass of data was so large that a computer was used to calculate the mean length at capture and mean back-calculated length at each age, with the variances and standard deviations of those figures. To analyse the effect of temperature on growth, the mean lengths for the first five years of life of females of five different year-classes were plotted against the cumulative number of degrees above 14°C, from water temperature records taken daily. The results showed that in general the higher was the cumulative number the greater was the mean length of the year-class. That is to say, a definite connexion was demonstrated between temperature and growth. How is it that the growth of pike in Windermere varies in different years according to the temperature experienced during the growing season? At least two factors may be involved. At higher temperatures the pike will digest its food more quickly and be ready to eat again sooner. This would be a direct effect. But in the better summers there is probably better light penetration into the water for longer periods of time—and the pike hunts by sight. Better sighting over longer periods could well lead to more predation. This would be an indirect effect.

In considering the combined effects of sex difference, inherent capacity to be a slow or fast grower, and weather conditions, the following comparison has been made. A fast growing female which grew in good weather conditions could measure 79 cm at age 4, whereas a slow growing male which grew in poor weather might measure only 54 cm at the same age (♀ weight 5200 g : ♂ weight 1400 g). The 25 cm difference between the two would be apportioned as follows: 8 cm due to sex difference, 12 cm due to fast or slow grower, 5 cm due to good or bad weather conditions.

The detailed discussion of the life-history of a coarse fish, the pike, has been used as an introduction to some of the basic conceptions and methods in fish biology, from which data are derived to explain past events and even predict future ones. The examination of feeding habits and age in relation to growth, and a consideration of the possible effect of a systematic fishing operation on the population structure, are always important and some translation of the information given can be made to other species by the reader. However, it might be as well to bear in mind that not all coarse fish are piscivorous like the pike, and many subsist on a diet consisting mainly of small invertebrate animals.

After making this reservation in exemplifying the coarse fish by the pike, we now move to a consideration of the salmonid fish, as exemplified by the salmon, the trout and the char.

Salmonid Fish

Although piscivorous and even cannibalistic individuals are not unknown within the salmonid species, the most usual feeding pattern is that aquatic invertebrates, such as crustaceans, molluscs and insects, and in some species small zooplankton, are taken throughout the whole of the life-history. This life-history is often so different from that discussed above that some elaboration is essential. Salmonid fish are often anadromous, that is they spend only part of their life in fresh water and the remainder in the sea. The brown trout is not normally anadromous, residing entirely in fresh water, but it may become so and is then known as a sea trout, but a single species definition, *Salmo trutta*, covers both types of individual.

Salmon

The Atlantic and the Pacific salmon both have a somewhat similar life-history. The fish spawn in freshwater streams and the resultant fry develop into fingerlings which spend from one to three years growing in fresh water. At this point migration occurs to oceanic feeding grounds, which are occupied from one to three or four years. The return migration back into fresh water then ensues, finally resulting in the sexually mature fish arriving at the site where they themselves were spawned.

Although if the fish is considered as an economic article the oceanic phase is the more important in that the main body weight is put on there, the freshwater phase is crucial and includes one of the most fascinating and still incompletely explained phenomena in the whole of biology, namely, the finding and recognition of the original natal (or parental) stream at the end of the return migration.

The story of the Atlantic salmon is an unhappy one. Whereas there were plentiful runs of the salmon up the major rivers of Western Europe until the last century, accelerating pollution has now eliminated most of these entirely. The River Rhine and the River Thames both had excellent salmon; paintings by the old masters show inhabitants of the Netherlands consuming fresh salmon as a normal article of the diet. Alas, these are no more. However, on the credit side a number of rivers in the British Isles still do have salmon and the Irish and Scottish runs are particularly strong, and zealously guarded. The greatest current danger to their continuance is posed by heavy gill netting off the shores of Greenland. This has acclerated tremendously over the last two decades, largely because of new knowledge that this area is a major gathering and feeding ground for the Atlantic salmon before they return to spawn. Atlantic salmon which spend only one year at sea before returning are called grilse and it is suspected that their major feeding area is closer to the British Isles than Greenland. So far it has not been discovered.

The Pacific Salmon

The Pacific Salmon is a in a stronger position. Alerted to the importance of the freshwater environment in the life-history and armed with an increasing body of research data, the American, Canadian and Russian authorities, particularly the Fisheries Research Board of Canada, have legislated to protect their precious charge. Of the seven Pacific salmon species the sockeye salmon, *Onchorhynchus nerka* (Figure 4.1 on page 61) has been studied the most intensively and some of its characteristics will now be discussed.[4] The sockeye has a freshwater distribution which extends from Kamchatka in the West, through Alaska to the Canadian seaboard in the East. For spawning, coastal streams or tributaries of major river systems which originate in lakes are usually selected. In the Fraser River, Canada, the returning migrants are 47-65 cm long and have usually spent two years at sea. The location of the natural spawning areas ranges all the way from only 50 miles from the mouth of this river to more than 700 miles upstream. Spawning occurs in August to November and the eggs are deposited in a redd, or nest, excavated from suitable gravel at the stream bottom. In making the redd, the female excavates a small pocket which is 20-30 inches across and extends 6-9 inches below the gravel surface. The fish pair to spawn (Plate 5) and the eggs and milt are liberated simultaneously over the centre of the pocket, into which they fall. The female then lays further successive batches of eggs in a number of pockets, always in an upstream direction, and since the disturbed gravel is directed downstream by the water currents, successive coverings also take place. The completed system of pockets constitutes a single redd. Although the sex ratio of the fish in the sea is 1 : 1 or even slightly in favour of the females, it is well substantiated that the gill nets which are in common use against the sockeye in the coastal areas and estuaries as it runs in to spawn take a larger proportion of the males. This is because of a sexual dimorphism in the fish at maturity. The males become hump-backed with a hooked snout and so are caught more easily. In both sexes the spawning colours are bright red overall, except for an olive green head; this is a complete transformation from the bright silvery hue of the fish in the sea. Despite the biased cropping of the males in these fishery operations (the sockeye makes excellent eating), direct close observation has shown that an excess of females on the redds is not fatal, even though consistent partnerships there are usual. A single male may fertilize the eggs of several females with no loss of viability in the eggs.

In order to develop in the redd the eggs must remain in the dark and be aerated by an adequate water flow during the incubation period of 80-140 days. After their emergence from the gravel in the early spring, the fry migrate to their adjacent lake and remain there for 1, 2 or 3 years, during which time they are known as fingerlings. The fingerlings feed on zooplankton, with *Bosmina, Cyclops* and *Daphnia* particularly favoured, and there is some evidence that these specimens are captured individually, rather than collectively, by merely straining through the gills and gill rakers. During this time zooplankton may be followed in its diurnal vertical migrations as it rises to the lake surface during the night and subsides at dawn. The larvae of chironomid midges and the larvae and pupae of other aquatic insects also figure in the diet. The rate of fingerling growth, determined by the availability of suitable food, has a great influence on subsequent events. In the rivers of British Columbia, Canada, the fingerlings usually migrate to the sea as smolts when they are 6-10 cm in length and have spent one year in the lake. The evidence is that only the few fingerlings which grow tardily and are not sufficiently developed to migrate at 1 year do so at year 2. On the other hand, the fingerlings in lakes in Alaska and Kamchatka often stay for two years or even three in some areas. In Lake Dalnee, Kamchatka, it has been suggested that although a two-year stay is generally attributed to shortage of food, in some years it may actually be attributed to *excess* of food. In these years the food resources of the lake are not fully taken up and the fish, staying on to feed, fail to develop the normal 'migrating disposition'. The latter is essentially a physiological state in the potential migrants which results from increased pituitary-thyroid activity induced by increasing hours of daylight (photoperiodism). As the season advances, thyroid activity decreases and a migratory urge is no longer possible. The scientific interest in the rate of growth of sockeye fingerlings in their lakes and their period of residence there is not entirely an academic one. The evidence from the Canadian studies is that the larger is the mean length of the year-1 smolt at the time of its seaward migration (comparing different yearclasses), the higher is the percentage return of adult fish. When the mean smolt weight increased from 4 to 10 g, the mean percentage survival tripled. This suggests that there is better survival not only during the actual migration to the ocean but also in the years of oceanic residence. So clearly, if the river system is to be the basis of an economic and thriving fishery this is an important consideration in its management. The Canadian view of fingerlings of two year lake residence is that these are undesirable from the point of view of fishery management. They compete for food with the fingerlings of the subsequent brood year and may even stay on as a 'residual' population with a capacity to come to maturity without ever leaving the lake. If this happens the case against them is heavy since they are known to be active predators against the sockeye fry!

In the early years of research on the sockeye salmon of the Fraser River, it was noted that, in nature, although a reasonable percentage of eggs was fertilized the losses of these were very heavy and the resultant hatch of fry on a 'percentage of eggs' basis was very poor numerically. On the other hand it was known that by hand-stripping the ripe eggs and milt from the fish and by using hatchery facilities for the subsequent development, a very high percentage of fertilization and production of young fish could be assured. Therefore it was argued that the hatchery incubation and subsequent release could augment a lake stock of fertilized eggs, fry and fingerlings as required, to utilize the feeding resources of the lake more fully and eventually give a higher production of smolts and adult fish. Five hatcheries were established for this purpose, but after some years of operation it was suspected that no benefit was accruing and accordingly rigorous trials were conducted. At Cultus Lake all the ascending sockeye were stopped at a weir. In the years when the natural spawning was allowed the fish were merely counted through the weir and allowed to proceed to the lake. At Cultus the sockeye population is unusual in that it spawns in the shallows of the lake rather than in the inflow stream. In the years of artificial propagation the fish were held below the weir until fully mature and then stripped by hand in the hatchery. The numbers of eggs reared to the eyed stage and planted or the numbers of fry hatched and released were estimated, as were the numbers in the years of natural spawning. The success of the propagation by both methods was determined by counting the number of seaward migrants. In eight years of natural propagation 1·0-4·2 per cent of the eggs became migrating smolts; in five years of artificial propagation 1·0-3·6 per cent of the eggs became migrating smolts. Therefore there was no significant difference in the yields and it was concluded that whatever advantage the hatchery system may have had in incubating eggs and producing fry these advantages were subsequently lost on release. Heavy predation in the lake was clearly responsible for the result. Tests were also conducted on pond retention to determine the merits of release at a later stage in development when the fish might be expected to avoid predation more expertly. This was shown to occur but the economics moved into an unsatisfactory state because of the higher costs of longer retention and the greater risk of disease epidemics occurring in these older fish. Attention having been drawn to the losses of sockeye fry and fingerlings by the activity of predatory fish, the latter were systematically removed from Cultus Lake by netting over a three year period. A three-fold increase in the percentage of eggs becoming migrating smolts was recorded, from 3·1 to 9·9 per cent, and this was attributed directly to the removal of the predators rather than, indirectly, to removal of competition in zooplankton feeding.

In the Pacific Ocean the sockeye smolts feed on pelagic plankton, particularly Crustacea, including the important copepod *Calanus*, which contributes so much to the fertility of the marine environment. Pelagic tunicates such as *Oikopleura* are also taken, and even squids and exceptionally small fish have been found in sockeye stomachs. Several vast common feeding grounds are recognized in the North Pacific and stocks from several origins mingle in each of these. From an area near the Aleutian Islands, for example, tagging has shown that the fish eventually travel in three different directions, namely north, east and west on their return migrations. Fish from the western Canadian seaboard seem to be confined to the Gulf of Alaska and mingle there with some of the Alaskan stocks. Until recently fishing was largely confined to coastal and estuarine waters as the sockeye moved in to spawn. To a large extent this coastal fishing could be associated with the spawning runs to adjacent river systems and regulated to allow a sufficient number of fish, the 'escapement' to return to spawn. As in the case of the Atlantic salmon however, the recent development of high-seas fisheries with motherships and attendant catcher boats has introduced a worrying new factor. Already it is known that the runs of sockeye to Alaska and Kamchatka have declined. It therefore becomes of vital importance to identify the source of origin of oceanic stocks to ensure that there is a proper balance and regulation of this fishing. Attempts have been made to investigate a serological differentiation of racial identity and some progress has been made in this direction. The most promising avenue investigated so far, however, is body scale examination. In sockeye of the Fraser River it was noticed that in the area of body scale laid down during the freshwater residence, counts of the circuli and an analysis of their nature (straight, crooked, broken, etc.) related particular scale types to different nursery lakes. To make the subsequent identification of the adult fish more exact, and since the circulus number varies to some extent from year to year, it is necessary to sample the seaward migration of the smolts in each year. This work has reached a level of sophistication where plastic impressions of the scales are used for the examination. These have the advantage that duplicates can be prepared for convenient storage and exchange with other scientists. By this means the fishing effort on the Fraser River can be regulated in relation to the individual spawning groups as they come in from the sea, each destined for a different spawning area. Close control of the whole operation is obtained easily. Theoretically the same system should give the required information for oceanic migrations and ranges and much progress has already been made in this direction.

Assuming that all the required information is to hand on the identity and numerical strength of the fish returning to spawn, the question which arises is what proportion of the total can be removed without jeopardizing the maintenance of the stock. In other words, what catch to escapement ratio will provide the best production. This is a very difficult question to answer. In the first place the annual variations in production, indicated by the number of adult fish returning, are quite marked. For example, in the Fraser River the mean returns per female spawner on the redds have ranged from 2·2 to 15·1 in different years. It seems that this variability from year to year is a reflection of what happens in the freshwater developmental stages, since there is a close correlation between the numbers of seaward migrating smolts in any one year and the numbers of adults derived from these which eventually return (Figure 4.12).[5] Therefore, it will be difficult to arrange the escapement to suit the weather conditions in the subsequent summer, for this will probably be the main determinant in this annual variation in production. A certain flair in prediction, possibly aided by the arts of astrology, would be useful for a sockeye biologist! There is accumulating evidence that the escapement must not

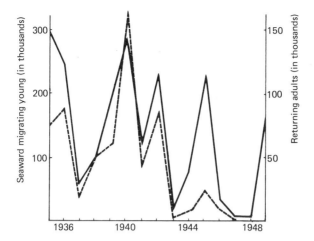

Figure 4.12 Relation of number of seaward migrating sockeye salmon (solid line) to number of adult fish returning (broken line) to Lake Dalnee, Kamchatka. Year dates are those of time of spawning, i.e. of the brood years (from data of Krogius, 1951)

be cut down to such an extent that nature is totally deprived of her apparently wasteful and profligate methods. After the sockeye spawns, every individual dies by natural ageing processes and the decomposing carcasses release nutrients into the water; these nutrients markedly enrich the spawning sites and nursery lakes and eventually by working through the food cycles nourish the young of the next generation. If the escapement can no longer provide this nutrient enrichment at a suitable level then an apparently irreversible cycle of declining yields is set in chain, measured by a continued fall in return per spawner. This appears to be happening in certain Alaskan rivers at present. Apparently the balance of natural predation is disturbed and under these circumstances the fry are attacked excessively.

For all seven species of Pacific salmon the evidence is that the adult fish return to the parental stream to spawn, or if the eggs are transferred to a different area for hatching and liberation they return to this new area. We may note here that precision of homing may be less in other salmonid species, for example in the sea trout it is well known that the fish can run to spawn in an adjacent river rather than the parent one. It is generally agreed that the anadromous salmonids do not feed during their return migration to spawn in fresh water. An interesting experiment began in 1935 on the sockeye salmon. Plantings of fertilized eggs were made in specially prepared redds in Spring Creek, a tributary of Cultus Lake on the Fraser River system which had never been frequented by spawning sockeye in previous years. This was almost certainly because the stream bed was composed of hard compacted shale, unsuitable for egg deposition, and it was therefore dug up in the experiment to receive the experimental plantings. Three years later, in 1938, more than two thousand adults returned to Spring Creek to find that the stream bed had reverted to its compacted state and was again largely unsuitable for egg deposition. Many fish died unspawned and in subsequent years Spring Creek again reverted to a largely barren arena. The mechanism of direction-finding in the run to the parental stream has interested scientists for many years. A solar orientation and sensitivity to a salinity gradient may give general directions for movement until the home estuary is reached. Once in the correct river system the fish may be quiescent for certain periods, but sudden increases in the water flow stimulate the ascent. Such incursions of increased flow are termed 'freshets' and are normally caused by heavy rainstorms in the river watershed. Migrations during freshets may be by day or night, suggesting that light is not an important trigger here. Guiding factors which may be of general importance in the migration run are temperature gradients and even carbon dioxide levels in the water, but eventually the final question must be posed. How is the exact parental site identified? On the Fraser River many spawning groups are en route, more or less simultaneously, each bound for different destinations. How can these groups maintain and perpetuate their collective identity? For the sockeye salmon there is no information or hypothesis, but recent fascinating work by Hans Nordeng on the migratory Norwegian char (*Salvelinus alpinus*) may be apposite here.[6] In Norway, smolts of the migratory char descend to the sea when 3-4 years old and return to their native river after only one month in coastal waters. From then on each individual migrates to the sea once every summer during the rest of its life. Nordeng studied the char of two adjacent river systems discharging into adjoining estuaries, the Løksebotn and the Salangen in Northern Norway. Fertilized eggs from the Salangen river system were transported by air to South Norway and hatched and reared in the hatchery at Voss. Four years later these fish were tagged for identification and transported back to the Løksebotn and Salangen estuaries. Of 174 released in the Løksebotn estuary 31 were recovered, 10 in the Løksebotn river and 21 in the Salangen river. Of 143 released in the Salangen estuary 26 were recovered in the Salangen river and only one in the Løksebotn river. These results suggest that the released fish preferentially returned to the river of their parents, lured by an attractant substance released by their relatives both old and young, still residing in that river. There was another piece of evidence tending to confirm this hypothesis. Offspring of the migrating population from Salangen were reared for three years in the Voss hatchery. In the spring, 200 immature individuals of smolt size were tagged and released in a lake below the hatchery. A fortnight later many of these fish were observed under the floor of the hatchery, positioned at the outlets from the tanks containing the rest of the population, comprising 2000 fish. Individuals from the large population of local non-migratory char were not observed there.

Nordeng cited another instance where a substance released by relatives may be acting as an attractant. In 1957 the Russians started to introduce pink salmon (*Onchorhynchus gorbuscha*) into rivers of the Kola peninsula, and in 1960 thousands of mature individuals were caught along the Norwegian coast and around Iceland and Scotland. Presumably these fish did not succeed in returning to the Kola rivers. However, with continued release of pink salmon fry in the Kola rivers and directly into the coastal waters, catches of mature specimens in Norwegian waters decreased to zero and a naturally spawning population became established in the Kola rivers. It is assumed that the build-up of the resident population eventually led to the recognition of the site on which to home. The Nordeng hypothesis allows for the recognition of an attractant from a native population into which the fish are introduced, in addition to the attractant from their own immediate relatives. When fish return from the sea their behaviour will be determined by the dominant attractant. In the char experiment the latter attractant was operative, being the only one present. In the Cultus Lake experiment described above, and in other cases where homing has been to the site of a new parental stream, after transfer of the eggs, the former attractant has been dominant. The experimental plantings at Cultus were derived from local fish, from Cultus Lake in fact, and so only the very final portion of their journey remains entirely unexplained. As in many experiments of this kind, particularly those which have been very thoroughly documented, the acquisition of further knowledge can often lead to a useful re-examination and re-interpretation of the result observed.

Further Aspects of Fish Feeding

The feeding patterns of pike and sockeye salmon have been discussed in some detail in the preceding sections. Both of these fish develop in lakes and it remains to consider the special feeding habits of fish such as the brown trout which may spend their whole lives in streams or rivers. It may come as a surprise to find that resident fish in flowing waters exhibit some degree of territorial behaviour, characteristically facing up-stream when quiescent but darting away to secure food and then returning to their former position. This behaviour can be observed from the bank on a suitable calm day. The stocking of a river with fish has to take account of the amount of food available on one 'fish beat' and overstocking can result in the production of small individuals feeding at starvation levels. In a Lake District study of young brown trout fry in two becks, it was found that aquatic stages of insects, especially larvae of the Chironomidae (midges) and nymphs of the Ephemeroptera (mayflies) were taken in both, while the water shrimp *Gammarus* was a large food item in the beck where it was abundant.[7] As the fish grew older, with an age of up to one year, the feeding pattern did not change significantly, although in the beck containing *Gammarus* this food was supplemented with animals of terrestrial origin, especially earthworms and insects. The characteristic rise of a trout to snap at a fly will be remembered in this connection. But caterpillars and even spiders and beetles figured in the diet at this stage! With the fish over one year in age there was still no marked change in diet, although the larvae of Trichoptera (caddis flies) were now consumed freely.

In this study the predilection of the fish for mayfly nymphs, especially those of *Baetis*, was particularly noted. These nymphs browse on the algae and detritus on stones in the stream bed but are easily dislodged and drift along in the water, although they soon recover their former postion as turbulence brings them back into contact with the stones once more a little further downstream. The trout evidently feed on these nymphs in the 'drift', where they are fully exposed to predation for a short while. The most interesting aspect of the drift behaviour of mayfly (and stonefly) nymphs in relation to trout feeding is that it is more pronounced at dusk and in the hours of darkness than during daylight hours. It is greatly reduced during nights of bright moonlight. These observations are explicable on the basis that the insects show negative phototaxis and move out from a position below the stones, maintained during the day, to a position on the upper surfaces of the stones during the night, providing the night is dark. The quick shining of a torch down into the water can reveal them at this time. Clearly the upper surface position on the stones is more desirable from the point of view of algal browsing. Although only a small proportion of these animals are in the 'drift' at any one time, compared with the total population in the whole stream, they constitute a major source of trout food. Whether or not the trout feed on them in complete darkness is not resolved at present, but the dusk feeding period is well established. In a stream in the Pyrenees (in a situation with no *Gammarus*) it was shown that feeding at dusk on drift from the benthic invertebrates, especially *Baetis*, was the sole means of trout sustenance in the early summer. In August, however, there was a second feeding period during the daylight hours when certain emerging aquatic midge insects and some terrestrial insects were taken. Dusk feeding was sufficient for resting metabolism but the supplement of the 'second meal' made growth possible.[8]

A final feeding pattern to be considered is extra-ordinary in that the mature fish, *Tilapia nilotica*, has successfully exploited a situation where small algae in the plankton are plentiful throughout the year and these are consumed as the sole diet.[9] *Tilapia nilotica*, (a coarse fish of the family Cichlidae) occurs in Lake George, Uganda, where it is an important part of the commercial fisheries. The feeding of the young fry in the littoral areas is not unduly remarkable since a wide variety of plant and animal materials are consumed including algae, detritus, rotifers, copepods, water mites and insects. However, as the fish grow larger they move offshore and an increasing proportion of planktonic algae is consumed, after they have attained a length of 6 cm the algae are eaten exclusively. Feeding is by a gulping action which resembles exaggerated respiratory movements. The plankton of the lake is dominated throughout the year by blue-green algae (see Chapter 2) and of these *Microcystis* is taken preferentially by the fish with some *Anabaena* also. There is a strong diurnal pattern in the feeding, and digestion relates to the condition of the stomach at particular times of the twenty four hour cycle. Feeding begins at dawn, when the empty and contracted stomach has digestive juices which are not strongly acid (pH 7·0). At first *Microcystis* tends to by-pass the inner recesses of the stomach and passes through the gut largely undigested. As the stomach fills, however, and increasingly acid digestive juice is produced (pH 2·0 or less), this reduces the *Microcystis* to a soluble form which can be absorbed. Feeding ceases at sunset and the stomach empties through the night. The success of this fish is demonstrated by the production statistics. Of 5000 tonnes of fish removed from Lake George annually for food, 80 per cent is *Tilapia nilotica*.

Fish Introductions

The introduction of fish into new freshwater environments, previously outside the natural range of distribution, has gone on for many years. In most cases the motivation has been to provide a thriving source of a desirable fish for sport fishing or human consumption, or both! Brown trout were introduced into New Zealand successfully and an attempt was even made to establish an Atlantic salmon stock. In the latter case however, although establishment was made in the Te Anau River the fish did not develop the habit of migrating to sea. However, other introductions have succeeded in establishing a migratory stock. The Russian work in the Kola peninsula has been mentioned previously. In the Falkland Islands, brown trout introduced from England unexpectedly developed a seaward urge and the examination of scales indicated that although some were 'slob fish', migrating only to the estuaries, others spent some time in the true marine environment. This success has encouraged the attempt at another Atlantic salmon introduction, it being argued that the Falkland Islands climate might be more amenable to the natural development and life-cycle of this species than that of New Zealand. The introduction of rainbow trout (*Salmo gairdneri*) of American origin has proceeded widely in European rivers and reservoirs. Although the fish has a high growth rate, reaches a good size, and has excellent eating qualities it has the disadvantage that it rarely breeds naturally in Europe; for example, in Great Britain and Ireland, of 550 waters holding rainbows only 5 contain self-perpetuating populations. Constant re-stocking of most of these waters is therefore necessary. In the European commercial hatcheries, mature rainbow trout are stripped of their eggs and milt by hand, to obtain fertilized eggs; these are grown up to a suitable size for subsequent release into sport or commercial fisheries.

An interesting coarse fish introduction in recent times has been that of the pike-perch (*Sander lucioperca*). After evidence indicated that it had some capacity to colonize fresh habitats in France, starting from a source on the River Rhine, it was deliberately spread into many French rivers and ponds, especially from 1950 onwards. This fish has excellent eating qualities and grows well. It has apparently found a niche which it can maintain easily, particularly where former natural populations of perch have declined.

Another type of fish introduction which may prove beneficial is that of species which control aquatic weeds and excess algal growth. With the increasing chemical enrichment of inland waters due to the widespread use of soil fertilizers, and the influx of domestic wastes and sewage effluents, so also is weed nuisance increasing. In Malaya, the grass carp (*Ctenopharyngodon idella*), which is a native of the West River in China and the Amur River in Russia, has been used in clearing fertilized fish ponds of excess weed growth and the wide temperature tolerance of this species has suggested that it may be useful elsewhere. There is little danger of it running wild in the areas of introduction since it rarely spawns naturally outside its native localities. Elsewhere, spawning must be induced artificially by pituitary injections. The fish species which best control algal growth are members of the genus *Tilapia*, and these are all tropical. However, the silver carp is also quite efficient and this has a wider range. It has been noticed in France that the activity of the grass carp may enhance algal growth, possibly due to the alteration of the nitrogen equilibrium in the surface mud as mentioned below. The ideal combination would therefore seem to be grass carp plus silver carp, if this can be achieved. In view of the two million pounds spent annually by River Authorities and Drainage Boards on the clearing of aquatic weeds, the Ministry of Agriculture, Fisheries and Food are conducting trials on the introduction of grass carp into Britain.

Fish Ponds

In eastern European countries, Israel, and Malaya, the carp (*Cyprinus carpio*) is grown for food, and this cultivation usually proceeds in ponds rather than lakes. The ponds are not necessarily natural but may be constructed or enlarged by excavation. Because of the close control which may be exerted over ponds, the results of alternative stockings and treatments may be examined more closely in relation to the turn-over of materials than is usually possible in lakes. However, the varied nature of the feeding of even one species of fish is such that the passage of materials through the food web is very difficult to follow. Thus carp feed mostly on aquatic invertebrates, but may also feed on fish fry and eggs, and even on vegetable matter. It has been shown that the nature of the fish stock influences the nature and metabolism of the whole pond community. For example it is known that the species composition of the zooplankton is different when carp is grown alone as compared with the composition when other fish accompany it. From Israel it is reported that carp yields may increase if other fish accompany it, but this has not been a universal experience so far. It is customary to fertilize the ponds. The yield of carp finally achieved is a function of the production of algae (primary production) and hence the effect of fertilization extends right through the food web. There is a closer relationship, an almost linear one in fact, between the amount of nitrogen in the water and the yield of carp (Figure 4.13), and hence it seems important to keep

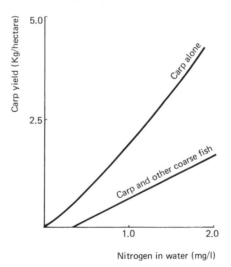

Figure 4.13 Relation between nitrogen in water and fish yield, in ponds in Czechoslovakia, (from data of Hrbáček, 1969)

the vital elements such as this in solution rather than immobilized in the bottom muds.[10] Since it has been shown in this work that the amount of nitrogen in only a 1 mm layer of the bottom mud is nearly as large as that in a 1 m deep layer of water above it, the dynamic equilibrium at this interface becomes important. If the equilibrium could be altered favourably, then the effect of fertilizer addition might become much greater; it is known that additions such as superphosphate have some effectiveness but become largely locked in the muds. It has actually been demonstrated that the carp themselves have some effect on the nitrogen equilibrium as the result of their burrowing in the mud for their food; when carp are removed from a pond the level of soluble nitrogen drops. Unexpectedly the yield of carp does not seem to bear any close relationship to the biomass of the benthic invertebrates present in the mud but it has been pointed out that fish yield is dependent on the benthic animals consumed, and in determining biomass of benthic animals one only estimates what is left behind. The two quantities are not necessarily closely connected; such a difficulty illustrates well the necessity for the study of dynamic rather than static parameters to examine fish productivity.

In addition to the use of ponds to produce marketable freshwater fish, attention has also been given to the possibility of altering the fish population of natural lakes to gain the same end. The Windermere operation to control the growth of pike was mentioned above, as was the Cultus Lake experiment in cropping the predators of the sockeye. The Russians and Americans are particularly interested in this aspect. Their biological measures include the very intensive catching of predator and non-commercial fish and even the total elimination of the resident population with ichthyocides, before re-stocking with carp and with whitefish (*Coregonus*) or other desirable species. Liming and fertilizer addition is also used to raise the level of productivity in the water. Research in this field studies the progress of the re-stocking balanced against the almost inevitable return of components of the original resident population.

The Introduction of Food Organisms

An increasingly modern trend in the usage of lakes is their utilization as storage reservoirs to provide domestic water or potential power for hydro-electric schemes. In northern Sweden most of the larger lakes are impounded and regulated for the latter purpose. This means they are dammed up above the natural water level in summer and autumn and lowered below the natural level in winter and spring. The effect in regard to fish is that certain food organisms previously present in the littoral bottom muds tend to disappear, particularly the important crustacean *Gammarus*. From the examination of stomach contents of resident fish, particularly char (*Salvelinus alpinus*), it was shown that after impoundment bottom-living animals figured less in the diet, *Gammarus* not being recorded at all, and more zooplankton had to be taken (Figure 4.14). Under these circumstances the growth rate of char decreased. In order to maintain and even increase fish production, the possibility of introducing new food organisms was considered, selecting those known to be less dependent on the spoiled littoral bottom zones. *Mysis relicta* is

86 Fish Biology

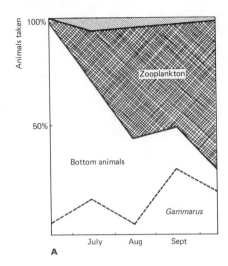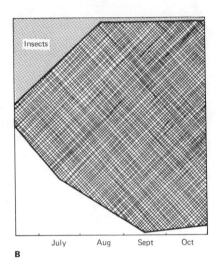

Figure 4.14 The feeding habits of char in Lake Blåsjön A before impoundment B after impoundment C after introduction of *Mysis relicta* (after Fürst, 1972)

such an acceptable food organism; it is a large crustacean (15 mm long) and only half-benthic, very mobile and with a vertical diurnal migration. In 1964, 1·65 million individuals were introduced into Lake Blåsjön. *Mysis relicta* lives on living and dead phytoplankton, dead zooplankton and bacteria, and the evidence was that the lake could sustain its growth. However, some years elapsed before the introduction was stabilized at a sufficiently high level to constitute significant fish food and it first appeared in stomach contents in 1968. In 1969, sampling showed that there were some *Mysis* at 0·5 m depth, a higher density at 1·5 m depth and up to ten to twenty individuals/m^2 at 10 m depth. Currently *Mysis relicta* is almost the sole food of the char in the months of October to June, the fish switching to zooplankton in July to September (Figure 4.14C).[11] Thus the char has gained compensation for the lost share of bottom animals and the results from the point of view of restoring the fish population appear to be most promising.

Mysis relicta has also been introduced successfully into Kootenay Lake in Canada. Here the objective was to provide a supplementary source of large invertebrates during the period when the resident rainbow trout switch from feeding on insects of terrestrial origin to preying heavily on other prized fish, especially the kokanee (a local non-migratory *Onchorhynchus nerka*), and depleting the stocks of these. The success of this particular introduction was assisted by increasing enrichment of the lake caused by an industrial outfall releasing phosphates. Under these circumstances *Diaptomus* and *Cyclops* previously present in the zooplankton have increased. The kokanee gains both from these early in its life and the *Mysis* at later stages, and is now suffering less from predation than it did formerly. It has held its own on a numerical basis and shows a higher growth rate.

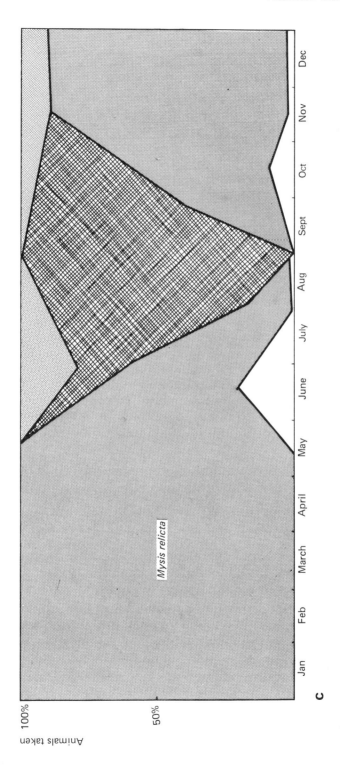

Heated Effluents

Another modern development which affects fish is the discharge of heated water from thermal power stations into certain lakes. An investigation in Poland showed that in such a lake their surface water temperatures were 8-11°C higher than normal in both summer and winter with a maximum of 32 °C and a minimum never less than 7 °C. At a depth of 10 m, 18 °C was recorded in the summer while at the same time and depth in a similar local lake with no heated water the temperature was only 8 °C. This is quite a different order of temperature elevation from that noted above in connexion with enhanced pike growth during the warm summers in Windermere. In the Polish lakes the effect of the heated effluent was not deleterious as might have been feared, in fact the reverse was the case.

None of the resident species, all coarse fish, were eliminated, and most showed an increase in their growth rate of from three to eight times as compared with a normal lake. This was ascribed to a more abundant zooplankton assisting in the sustenance of the young. The one reservation concerned the bream, which responded less than the other fish, and this was attributed to a slightly poorer bottom fauna in the heated lake. Pike showed a typical adaptation to the new conditions in that they spawned from late January until the end of March as compared with the period of March to April in normal local lakes. It was inferred that the pike had adjusted its annual cycle of development so that the temperature during spawning and egg growth did not exceed that normally experienced. Because of the known preference of salmonid fish for colder conditions the effects of heated effluents on them may be much more serious and their position must be under constant review in this connexion.

Serology

Fish serum is the clear component of the blood which carries the red blood corpuscles; it is easily removed from the latter by centrifugation. Analysis of the serum shows that many proteins are present and these can be separated in an orderly fashion using electrophoretic techniques. Here the different proteins each travel at different speeds in an electric field and for convenience the runs are often made through a thin starch gel and then stained for location and identification. Different running speeds are a function both of the different net charges on the molecules and of the different molecular sizes. It has been found that homologous proteins from different species or sub-species of fish travel at different rates and consequently this technique has an obvious potentiality for distinguishing racial stocks. This is particularly so in the case of the salmon where, although there is mixing of the stocks of these in the oceanic feeding grounds, there is also a rigid reproductive isolation maintained by the strong homing instinct. Atlantic salmon feeding together off Greenland have been shown by this method to be of two types, with sera patterns attributed either to European or to Canadian and North American stocks.[1,2] Such information is of crucial importance in relating the high seas fishing to conservation of the spawning runs back to fresh water, as mentioned above. Although serum patterns, so far, are distinguished on rather few differences, recent work has analysed the difficult region in the separation to which the transferrin molecules run. Transferrins are the proteins which bind ferrous ions in the serum. Such analyses may well prove to be productive of even finer race distinctions. It would be interesting to know, for example, if there is a serological difference, implying a genetical one, between salmon which enter their native British rivers in the spring and those which delay their entry until closer to the spawning time in the autumn.

Fish Diseases

Fish are subject to many different diseases, and these often become particularly apparent when a high population density is established. Hatcheries are an extreme example of this. The infections of fish caused by other animals, ranging from protozoa to cestodes and trematodes, are topics for specialist study and will not be discussed here. However, because infections by micro-organisms are of considerable economic importance, and in addition are often particularly dramatic and devastating, it is appropriate to consider these. Three important examples are described. These involve a virus, a bacterium and a fungus, respectively.

Viral haemorrhagic septicaemia

Viral haemorrhagic septicaemia, also know as Egtved disease or infectious anaemia, is a serious virus disease of rainbow trout; it has caused numerous losses in hatcheries on the European continent. The mortality is related to the age of the fish, fingerlings dying more quickly than older fish. However, the fertilized eggs appear to be invulnerable, even when spawned by infected fish, and this has had the fortunate result that the export of these from infected areas had not led to the establishment of new disease centres. So far Egtved disease has not been recorded in the British Isles, despite the presence of its host. In the acute phase of the disease the fish often show abnormal swimming movements and spiral repeatedly in the hatchery ponds. They also darken in colour and at autopsy haemorrhages are present throughout the viscera and muscles, and the liver is degenerated. The condition is spread by the urine and faeces of the fish, and it has even been suggested that gulls may be responsible for some of the transmissions, netting being used to exclude these birds in many hatcheries. Stress in the fish, induced by crowded conditions, often plays an important part in unmasking a latent infection. Because the outward signs and autopsy findings are not completely unequivocal for the virus, and in order to obtain a certain knowledge of its presence and possible route of transmission, a more exact diagnosis is made by two tests. The virus-isolation test is carried out by homogenizing the diseased trout tissues (the liver is an example), passing them through a fine-pored filter to eliminate bacteria, and inoculating the solutions into established tissue cultures prepared previously from healthy fish. If no virus is present the tissue cultures continue to grow normally. If Egtved virus is present the attack on the tissue culture cells is observed microscopically, although of course the virus particles themselves cannot be seen, unless an electron microscope is available. Fluorescent antibody staining of sections of diseased tissue is also used for diagnosis. In this test anti-Egtved serum is obtained by inoculating rabbits with weak strains of the virus and used to detect the presence of the active virus in the sections. These are treated with a suitable activator and the anti-serum is then applied. Where the virus is present a visible stained area of reaction is observed under the fluorescence microscope. The fluorescent-antibody test has the advantage over the virus-isolation test in that it is very specific for different strains of the virus, therefore it may give information on the possible source of the latter.

Strains of Egtved virus which had become non-pathogenic to fish, although still growing in tissue culture, were injected into healthy rainbow trout and it was shown that a weak antibody was produced as a result. However, it has so far not been possible to demonstrate that such an antibody is formed naturally in fish exposed to the disease. From the results obtained so far, therefore, it would appear that rainbow trout largely lack any natural resistance.[13] Complete disinfection of the hatchery ponds by liming is practised following bad outbreaks of Egtved virus, but in the long term the selection of strains of rainbow trout with a genetic resistance is desirable.

Furunculosis

Furunculosis is a highly infectious disease caused by a bacterium, *Aeromonas salmonicida*. Observations linking the bacterium with the disease were first made in Germany in 1894, but English reports followed in 1909 and subsequently there were others from Canada and the USA. Although salmonid fish, particularly brown and rainbow trout, are always thought of in connexion with this disease, it can cause mortality in many coarse fish species, including pike and even goldfish. In this and other fish diseases the salmonid fish have attracted the greater attention because of the higher economic value of the host. Furunculosis is probably native to western North America, where it had a balanced relationship with rainbow trout, a native fish of these regions. As is the case with other infectious diseases its introduction to new localities has disturbed this host-parasite balance, and when it reached the eastern United States severe epidemics in the brook trout (*Salvelinus fontinalis*) ensued. It is now endemic all over Europe but has not been reported from Australia nor New Zealand. Furunculosis is very contagious and is transmitted easily in water, especially if the fish are crowded, for example, in hatcheries. However, it has been noticed that the spread of the disease ceases if the fish are present in a quantity of less than five fish per two cubic metres of water. Since the bacterium is an obligate parasite in nature, it never occurs in water which has not been contaminated by diseased fish. Its survival capacity in the aquatic environment is low, only one week in unpolluted conditions.

Aeromonas salmonicida grows aerobically or as a facultative anaerobe and when cultured from infected tissue is highly pleomorphic, that is, the cells are of very varied sizes, and it produces characteristic melanin pigments on certain agars. These are the diagnostic features. After infection the bacterium invades the blood stream and internal organs. If the disease is acute, the fish may die within a few days with no outward signs, but with a less acute situation boils full of the bacterium form under the skin and in the muscles. The so-called furuncle nature of these boils gave the name to the disease. Open wounds may also appear on the body surface. It is known that fish may survive mild infections and a measure of this is the subsequent capacity of their blood serum to agglutinate living *Aeromonas salmonicida* cells. These fish have acquired a natural immunity, presumably as the result of antibody production. Some populations examined had a high percentage of individuals with a naturally acquired immunity. These observations encouraged the search for an agent with which fish could be immunized against the disease, and it was found that a preparation from disintegrated *Aeromonas salmonicida* cells was not toxic to the fish but did induce the antibody. This preparation was therefore fed to fish, incorporated into a pellet-form diet, and although full protection was conferred in some field trials others had only limited success. The mechanism of antibody production and induction by fish is not comparable to the mammalian one and is very poorly understood at present. Treatment and disease prevention in hatchery fish can also be achieved, fairly easily, by using sulphonamide drugs or antibiotics such as chloramphenicol or tetracycline as food additives, but there is always the danger with such treatments that resistant strains of the bacterium may evolve. This has already happened when certain sulphonamides have been used. Since the disease has been well-known for such a long time, breeding selection for resistant stocks has proceeded widely and resistant brook trout were propagated in certain parts of the USA as early as 1925. This is an aspect which will undoubtedly feature in future research work.

Saprolegnia disease

Saprolegnia disease of fish has recently caused great concern in the British Isles. The fungus responsible is discussed more fully in Chapter 6. On the fish it is visible as a spreading circular patch of white mycelium, and motile spores are released to act as the new infective agents. This disease has always been a problem in hatcheries where it is often controlled by using fungicides as additives to the water, but since 1966 it has come into particular prominence by attacking mature Atlantic salmon migrating into the rivers to spawn. Many of these have died before they could reach the spawning grounds. Trout and char have also been infected in the wild, but coarse fish are largely immune. *Saprolegnia* is not an obligate parasite and in nature grows in water as a general saprophyte on decaying animal material and even on dead vegetation. In view of this constant occurrence of the fungus in the environment, the recent outbreak of disease, termed salmon disease or ulcerative dermal necrosis, has been attributed by some to the arrival of a new pathogen, possibly a virus, with *Saprolegnia* acting as a secondary invader. Against this theory is the fact that no virus has been isolated so far and the observation that the *Saprolegnia* can be referred to a particular type of strain which seems to be largely restricted to fish for its growth substratum.[14] It is possible that especially virulent strains of this

Saprolegnia have evolved only recently. However, a study of this disease shows clearly that no single factor can be considered in isolation. For example, the physiological state and age of the susceptible fish are of extreme importance, generally only mature spawning fish are attacked. It is possible that a recent change in some environmental factor may have played a part in inducing the infections, although it is difficult to see what this could have been in Windermere, where the brown trout have suffered heavily.

References

1. W.E. FROST and C. KIPLING, 1967. *J. Anim. Ecol.* **36**, 651-93.
2. M. KENNEDY, 1969. *Irish Fisheries Investigations. Series A, No 5.*
3. W.E. FROST and C. KIPLING, 1959. *J. Cons. perm. int. Explor. Mer* **24**, 314-41.
4. R.E. FOERSTER, 1968. The sockeye salmon, *Oncorhynchus nerka*. *Fish. Res. Bd. Can. Bull.* **162**
5. F.V. KROGIUS, 1951. *Izvestiia* **TINRO 35**, 3-16.
6. H. NORDENG, 1971. *Nature, Lond.* **233**, 411-13.
7. J.C. McCORMACK, 1962. *J. Anim. Ecol.* **31**, 305-16.
8. J.M. ELLIOTT, 1973. *Oecologia* **12**, 329-47.
9. D.J.W. MORIARTY, J.P.E.C. DARLINGTON, I.G. DUNN, C.M. MORIARTY, and M.P. TREVLIN, 1973. *Proc. Roy, Soc. Lond. B.* **184**, 299-319.
10. J. HRBÁČEK, 1969. *Verh. Internat. Verein. Limnol.* **17**, 1069-81.
11. M. FÜRST, 1972. *Verh. Internat. Verein. Limnol.* **18**, 1114-21.
12. O.L. NYMAN and J.H.C. PIPPY, 1972. *J. Fish. Res. Bd. Can.* **29**, 179-85.
13. P.E. VESTERGÅRD JØRGENSEN, 1971. *J. Fish. Res. Bd. Can.* **28**, 875-7.
14. L.G. WILLOUGHBY, 1971. *Salm. Trout Mag.* **192**, 152-8.

Chapter 5
Streams and Rivers

In running waters, of streams and rivers, the nature of the biological productivity is determined by a number of factors which show obvious variation from place to place.[1,2]

Current speed and the nature of the stream or river bed are very influential. With the very swiftest flow, classified as torrential, across a substratum of rock or heavy shingle, the only plants which can maintain a footing are mosses, especially *Fontinalis*; this is securely attached by an efficient hold-fast. A slower current speed, again across an essentially stony bed, allows the establishment of some larger plants such as the water crowfoot (*Ranunculus fluitans*) and the river milfoil (*Myriophyllum spicatum*) (see Figure 2.9, page 39). With a still slower flow of water, silted conditions develop and, although *Ranunculus* is often still present, other submerged weeds such as *Potamogeton* come in. Finally, where the current is very slow, silting conditions give a deposition of mud and a varied community of other plants develops. These plants, such as the mace reed (*Typha*), the rushes (*Juncus*) and the sedges (*Carex*), form an essentially 'emergent' community as the plants stand up out of the water. In small streams with a muddy bottom, and where the water is fairly hard, the water cress (*Nasturtium officinale*) and the water parsnip (*Apium nodosum*) often occur.

In streams and rivers, the large plants, some of which have just been mentioned, would seem to constitute obvious foodstuffs for the invertebrate animals which are present. It is therefore slightly unexpected to find that this material is largely not utilized. When river weed is examined there is little sign of obvious damage from animal grazing, although there may be numerous insect nymphs, shrimps or snails clinging to its surface. On the other hand, vegetable materials of terrestrial origin deposited in streams, for example, leaves of trees such as alder, oak and sycamore, usually show clear signs of having been nibbled. When the water shrimp, *Gammarus pulex*, is washed out of oak leaves from streams it is more than coincidence that these leaves are often reduced to leaf-skeletons (Plate 6). However, river weed does have an importance in the invertebrate life-cycles in that it harbours attached (epiphytic) algal growth; the presence of this as a food material explains the occurrence of the small animals browsing amongst it. The situation is further complicated by the fact that the attached algae will also grow on the deposited terrestrial leaves. Thus invertebrate animals browsing amongst it have a mixed diet available for them (Plate 7) and in many instances the precise nature of this diet is a matter for controversy.

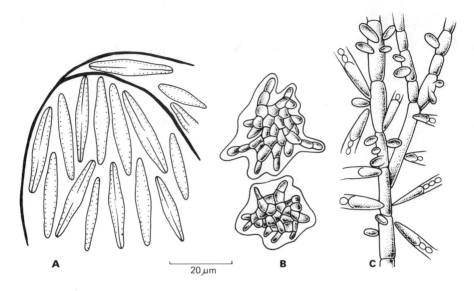

Figure 5.1 Stream algae **A** *Achnanthes minutissima* growing on stone **B** *Ulvella frequens* **C** *Chamaesiphon incrustans* growing on a filamentous alga. **A** and **B** are epilithic algae.

The importance of the attached algae in stream and river food webs having been established it is appropriate to consider them more closely. In the case of an oligotrophic or unproductive flowing water, with a low level of dissolved plant nutrients, diatoms such as *Achnanthes* (Figure 5.1) are dominant near the headwaters. This growth will occur entirely on rocks and stones if no weeds are present, and it has been shown that stone-grazing for this seemingly unpalatable food, which consists largely of silica shells, is carried out by *Agapetus* (see below).[3] *Achnanthes* has no capacity to resist desiccation and a receding water level, a not uncommon event in stony streams, will cause a temporary decline in the population as stones become exposed and dry out. As the water flows downstream it generally passes through softer rocks and gains dissolved plant nutrient substances from these and also from drainage and seepage. This gain in nutrients allows a more luxuriant attached algal community to develop and this is dominated by other diatoms, small green algae such as *Ulvella* and *Chamaesiphon* (Figure 5.1) and larger filamentous forms. *Chamaesiphon* often grows on the filamentous algae if these are present. In situations particularly rich in nutrients, the attached filamentous algae *Cladophora* and *Stigeoclonium* can develop into massive colonies. Again, these algae may themselves bear smaller ones on their surfaces (see Figure 2.12 on page 44). Running parallel to this change in the algae there is the greater probability of weed establishment, due to the slowing rate of water flow in lowland situations and the depositions of silt and mud. As stones diminish in importance as surfaces for algal growth downstream, weed surfaces replace them.

The invertebrate animals which live in the actual beds of streams and rivers, as distinct from those clinging to plant surfaces, have to find a living in situations varying from rocky and stony to muddy. It is not surprising to find that these two extreme types of situation favour distinct types of fauna. Rocky and stony, or 'eroding' substrata, associated with turbulent waters in upland situations, harbour animals which require a well-oxygenated and cool environment. Stonefly and mayfly nymphs, exemplified by *Amphinemura sulcicollis* and *Baetis rhodani* (Figure 5.2) respectively, are such animals. Stonefly nymphs can avoid exposure to the full force of the current, with the danger of being swept away, by creeping down amongst the stones. *Baetis rhodani* is flattened and applies itself closely to the stones, clinging on by its claws, and remaining passive while feeding. If dislodged however it is a furiously active swimmer and usually manages to regain another similar position. We saw in Chapter 4 that the habitual emergence of stonefly and mayfly nymphs at dusk onto the upper surfaces of stones to browse on algae often leads to their being swept away and becoming exposed to predation by fish at this time. This is especially so in the later nymph instars (up to a dozen may occur in the life-history from successive moults) when the animal attains a length of 3 mm or more.

Other and more obvious modifications associated with life in turbulent flowing waters are the construction of cases of heavy anchoring stones, by caddisfly larvae and pupae, such as those of *Agapetus*, and the development of large flat attachment feet, for example by *Limnaea pereger*, the only snail commonly found in such situations. A most fascinating exploitation of the fast-flowing environment is made by the net-spinning caddis worms. The animal constructs and attaches a net, the aperture of which is directed upstream and the force of the current keeps it open. The larva, which lives in the tail-end of the net, merely cleans its surface of algae and detritus at intervals.

When an upland stream is in full spate following heavy rains, there is the possibility that the entire fauna will be swept away. On the other hand, in times of drought, the stream may dry up so completely that it seems that the aquatic animals must perish through desiccation. It appears that survival is often possible through the very deep colonization of the stones in the stream-bed, where some water is always retained. In a study in Canada, Hynes drove perforated stone-filled pipes into stream beds to study the colonization.[4] A revolving outer sleeve, the perforations on which matched those of the pipe, allowed the apparatus to be first hammered into position closed – and then opened up. This precaution ensured that the animals were not carried down inside the apparatus. The results demonstrated colonization to at least 50 cm below the stream-bed surface, and a wide variety of different animals were present.

In the muddy or 'depositing' type of stream or river bed, the benthic algae which are present there are too capricious in their occurrence to be a regular source of food for the invertebrates. Rather it is the mud itself, and the decaying organic matter, such as dead leaves, and other detritus it contains, which sustains the animals. The generally sluggish current passes a less well-aerated water than that in the upland stony streams, and in hot weather, especially at night, the oxygen content may fall to a low level. The capacity to resist such a fall, and an even lower oxygen content in the mud itself, leads to colonization by a fauna which is similar to that found in purely stagnant conditions, as in ponds. The water-slater, *Asellus* (Figure 5.2) feeds on dead leaves and microbial growth at the mud surface. Below

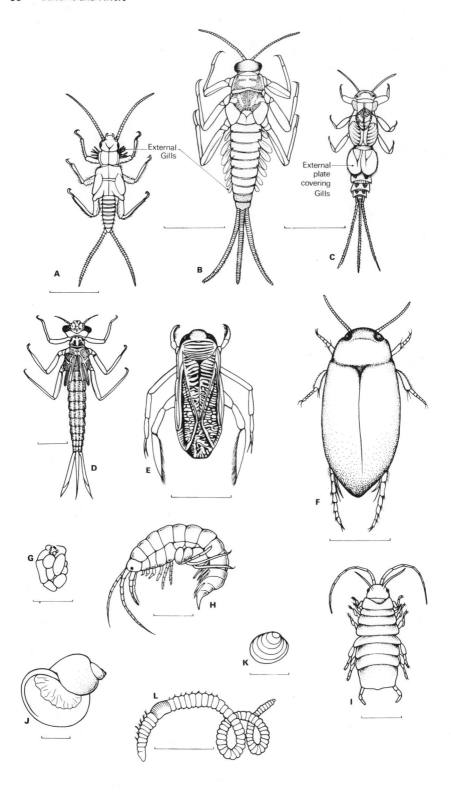

Plate 1

A Phytoplankton from a eutrophic lake (Esthwaite Water) mounted in dilute Indian ink to show mucilage coating the colonies of the blue-green algae *Anabaena circinalis* (N) and *Microcystis aeruginosa* (M). *Ceratium* (C) is also present. (Photograph by Hilda M. Lund.) **B** *Stigeoclonium* colony from a sewage effluent channel. (Note dense investment of filamentous bacteria which this alga tolerates.)

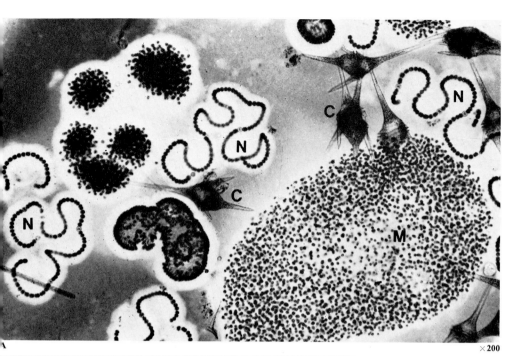

×200

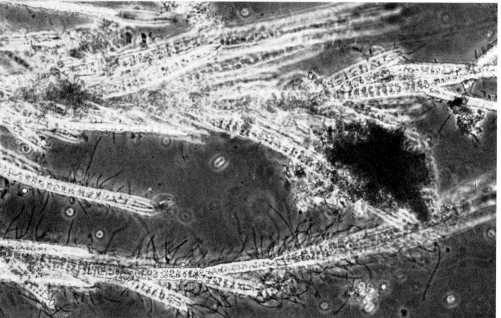

×500

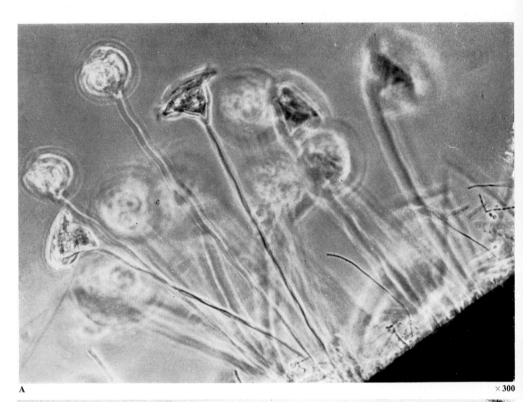

A ×300

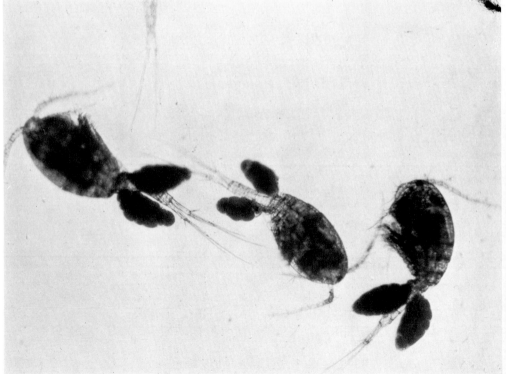

B ×54

Plate 2 *(opposite)*

A *Vorticella* feeding with active cilia   **B** living female *Cyclops* showing paired external egg-sacs

Plate 3 *(below)*

Pike, *Esox lucius*, opercular bone from an eight-year-old male fish: The first-formed annulus cannot be observed and only seven are visible, the last on the bone edge. (From Frost and Kipling, 1959.)

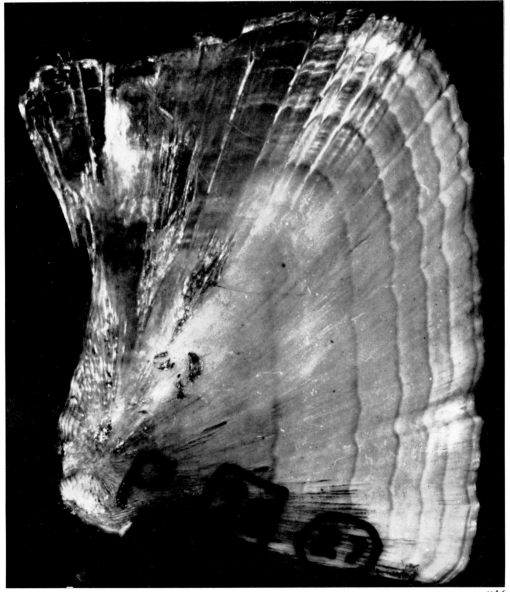

Plate 4 *(above)*

Pike, *Esox lucius*, approximate age two years and length 42 cm, consuming another fish. (Photograph by Oxford Scientific Films.)

Plate 5 *(below)*

Sockeye salmon, *Onchorhynchus nerka*, paired for spawning. Note also spent carcasses in right foreground. (Photograph supplied by The International Pacific Salmon Fisheries Commission.)

Plate 6

Gammarus pulex disturbed from a feeding situation under an oak leaf (note resultant skeletonization of this material) and showing postures varying between those of **A** active movement **B** immobility. (Electronic flash photographs by A. E. Ramsbottom.)

A   ×2·5

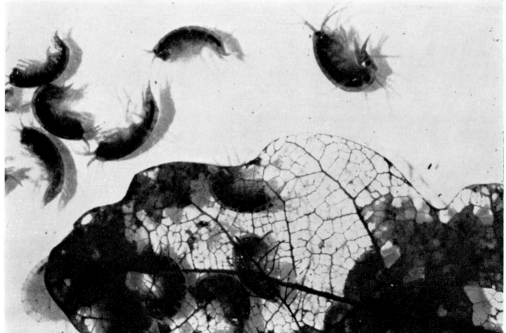

B   ×2·5

Plate 7

Investigation of the food of a stonefly nymph, *Amphinemura sulcicollis* (see Figure 5.2 A), in Mosedale Beck, Cumbria. At high altitude, situated above the tree line, the torrential flow prevents the establishment of rooted plants. The beck does however receive large amounts of grass blown in from the surrounding moor. This grass is colonized **A** by epiphytic algae which are **B** unicellular or **C** filamentous and **D** by fungi. **E** gut contents of the animal show that the grass (G) is consumed together with the filamentous algae (L) and presumably the fungi also. (All materials freshly collected from the site.)

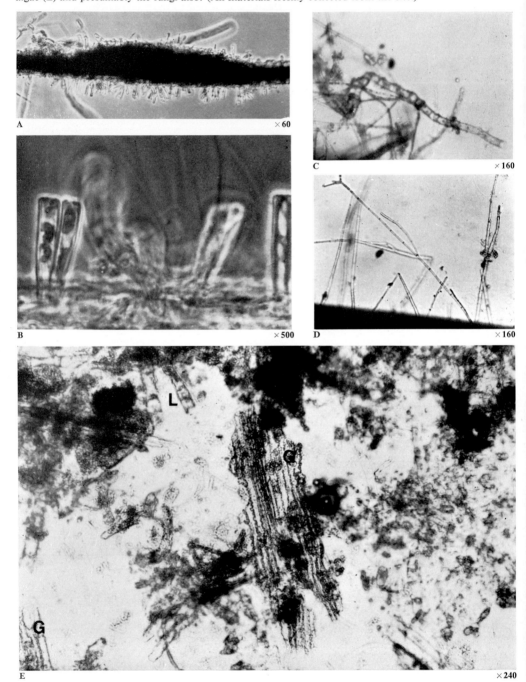

Plate 8

A *Saprolegnia* sporangium releasing motile zoospores. **B** Colloidal chitin-cycloheximide agar used for the isolation of aquatic actinomycetes from a neat stream water sample. Water sample overlaid and incubated. Note hydrolysis of chitin both by *Actinoplanes* colonies (arrowed) and some of the bacterial colonies. **C** *Actinoplanes* motile spore showing flagella. **D** *Sphaerotilus natans* filaments, with a diatom cell, *Navicula*, for size comparison. **E** Heterotrophic aquatic bacteria. (**A, D, E** phase contrast on living material **C** electron micrograph.)

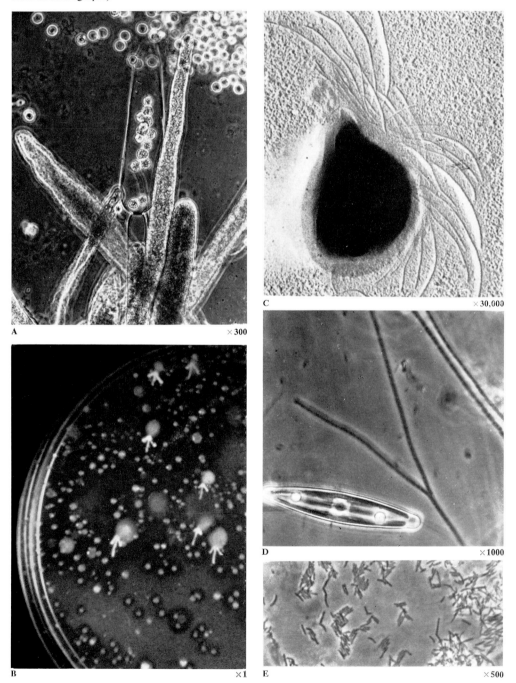

Plate 9 *(left)*

The dam at Cow Green Reservoir, Upper Teesdale, Durham, completed in 1970, to make a reservoir from a flooded upland valley system. This is a 'regulating reservoir' storing winter water and releasing it through the outflow stream (lower picture) during drought conditions. Abstraction for industrial purposes can now be made from lower sections of the River Tees throughout the year. (Photographs by A. E. Ramsbottom.)

Plate 10 *(right)*

Small-scale apparatus for testing the toxicity of poisons to fish (in the small flasks). (From HMSO 1972.)

this surface, and sometimes projecting into the water, are the Tubificidae worms, such as *Tubifex* (Figure 5.2), which collects what oxygen it can through the haemoglobin in its blood. Larvae of the insect midge *Chironomus* make tubes in the mud and drag saliva-coated debris from the mud surface down into them. The pea-mussel, *Pisidium*, lives just below the mud surface with part of the shell exposed; water and detritus is siphoned through the animal. In the 'depositing' type of stream where the silt is coarse rather than muddy other animals can occur, for example the mayfly nymph, *Caenis moesta* (Figure 5.2). In this nymph the external gills on each side of the body beat in such a way that a transverse current is produced, enabling it to respire in conditions which tend to be smothering. In addition, broad external plates cover the external gills and prevent the deposition of silt upon them. Here we see a difference from *Baetis rhodani*, the external gills of which are quite exposed and held rigid and immobile, aeration movements presumably being unnecessary in the turbulent streams which constitute its habitat.

The presence of plants on depositing substrata gives a habitat to many species which can tolerate the periodic occurrence of a low oxygen level but which cannot live in the mud itself. These include some of mayflies, dragonflies such as *Ishnura elegans* (Figure 5.2), caddis-worms which construct cases of vegetable matter, and green chironomid larvae. Some of the latter feed on the weed directly but others merely use it as an attachment site to filter out passing detritus. With a dense emergent flora, elements of a fauna (see Figure 5.2) more usually associated with ponds, such as the water boatman, *Corixa*, water beetles such as *Ilybius* and mites can come in.

So far, little has been said of the chemical nature of the water and it is necessary to consider the calcium content more closely. Few river animals are affected adversely by hard water, but those which form calcareous shells generally require at least 20 mg/l of calcium in the water and are more abundant when an even larger amount occurs. This is generally true for the snails and other molluscs although *Pisidium* can occur in very soft waters. A summary generalization is that stoneflies and mayflies are dominant in very soft waters; mayflies and chironomids are dominant in harder waters; *Gammarus*, oligochaete worms and snails are dominant in the hardest waters. The situation in regard to *Gammarus* and calcium can be quite complicated in marginal situations and the degree of water acidity has a bearing on this. In Britain *G. pulex* is not recorded where water pH is 5·8 or less. In a study of the River Duddon and its tributaries a whole series of pH values was recorded in the various streams, the variation depending on the nature of the ground rocks and modified by the addition of acidity (from sulphate) derived from smoke pollution and precipitated in rain. In the Crosby Gill tributary with calcareous water the pH was generally about 6·5-7·0 and *G. pulex* was present, whereas in the Gaitscale Gill tributary the pH was 4·5-5·2 and the animal was absent.[5] The significance of pH

Figure 5.2 Stream and river invertebrates **A** *Amphinemura sulcicollis* nymph **B** *Baetis rhodani* nymph **C** *Caenis moesta* nymph **D** *Ishnura elegans* nymph **E** *Corixa dorsalis* **F** *Ilybius fuluginosus* **G** *Agapetus* pupa **H** *Gammarus pulex* **I** *Asellus aquaticus* **J** *Limnaea pereger* **K** *Pisidium* **L** *Tubifex tubifex* (The scale line denotes 3 mm in each case) (some after Hynes, 1960)

rather than dissolved calcium as the determinant of the distribution of *G. pulex* is based on comparison with continental areas where the animal may occur in waters of pH 5·8 which have a fairly low calcium level. In such waters the input of acidity from the atmosphere is not a modifying factor in the water pH whereas in many British situations the incoming acidity must be buffered by larger amounts of $Ca(HCO_3)_2$ before a favourable background pH is achieved. It is very possible that the pH factor operates on *G.pulex* indirectly by affecting its food materials in some way and here we note the possible role of micro-organisms in the conditioning of this food.

Gammarus

Since *Gammarus*, the water shrimp, is such a conspicuous large crustacean of many streams and rivers more will be said about it here. *Gammarus pulex* is the species which occurs in flowing waters in Britain and another species, *G. lacustris*, is present in the static ones. *Gammarus pulex* may be collected on decomposing tree leaves in a fairly superficial position, as mentioned above, but in addition it may also be obtained by kicking up the bottom mud and stones and using a fine-meshed net on the cloud of debris which results. Presumably the finer particles of organic matter constitute its diet in this deeper situation. Submerged vegetation also harbours the animal and the aquatic moss, *Fontinalis*, can be particularly productive if it is removed and washed. At maturity *G.pulex* attains a length of about 13 mm, achieved through a series of moults made every two or three weeks by the juvenile stages (instars) which are essentially similar to the adults, but smaller in size. The mature animal swims upright or on its side, using both thoracic and abdominal limbs for its propulsion. Respiration is by gills at the base of the first four pairs of thoracic limbs. Reproduction is initiated when the male clasps the female and the pair swim together for several days until the female moults. Copulation then occurs and the male places sperm on the ventral surface of his partner, who then breaks away. Eggs from the ovaries are laid in a brood pouch before the cuticle hardens and fertilization occurs.

After some days, hatching of the eggs takes place and after a further period of a day or so the young leave the mother's pouch. The female becomes attractive to the male again at, or slightly before, the hatching of the eggs and pairing occurs again at this time. Several batches of eggs are laid and because the female is growing continually the number of offspring produced increases from six to about sixteen in successive broods. In Britain animals born in March will be sexually mature and breeding in July, August and September. During the winter, breeding ceases but the animals survive and produce spring broods before dying in May. On the other hand animals born late in the summer season overwinter as juveniles, come to sexual maturity in the following spring and die in the summer.[6]

Gammarus pulex can be an important fish food (see Chapter 4) and from the pattern of the life-cycle (Figure 5.3) it will be seen that some large adults occur throughout the year. The ones which overwinter are particularly useful to the fish since other animal foods (e.g. insect nymphs) are often less available at this time.

Freshwater Biology 99

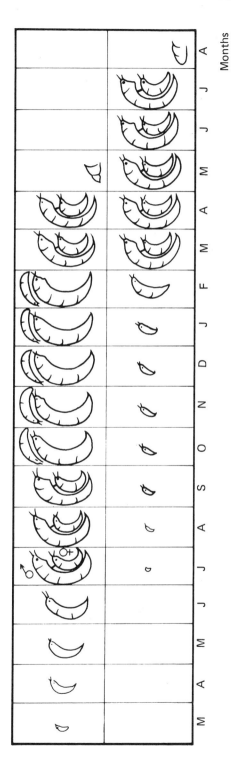

Figure 5.3 The life-cycle of *Gammarus pulex* in Britain showing that the mature animals may be collected in every month of the year

Sampling the Invertebrate Fauna

In sampling for the invertebrate fauna in streams and rivers the seasonal occurrence of some of the members necessitates several collections made in different months before a complete picture can emerge. Using the simplest method, of stirring up the river bed and netting any creatures which are dislodged, many of the smaller developmental stages are undoubtedly lost. However, such a method has shown that some of the stonefly nymphs, such as those of *Amphinemura sulcicollis,* disappear during the summer while mayfly nymphs, such as those of *Baetis,* can occur throughout the year (Figure 5.4). The apparent disappearance of *A. sulcicollis* is explained by the intolerance of the feeding aquatic stages to warm summer temperatures; these are evaded by enduring this period in the egg stage, laid by the adult insect in the summer.

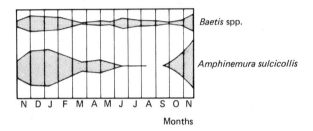

Figure 5.4 The numbers of two animals caught month by month in a stream using a standardized netting technique. In the kite diagram the width is proportional to the number of specimens of that animal caught (from Hynes, 1960)

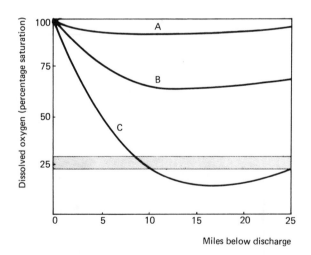

Figure 5.5 Predicted concentration of dissolved oxygen in rivers with the same flow of water and the same initial concentration of oxidizable matter. River depth (feet): **A** 2; **B** 5; **C** 10. Current speed (mph): **A** 0.5; **B** 0.2; **C** 0.1. The critical oxygen level, below which trout are asphyxiated, is shaded (from Southgate, 1969)

River Quality

Rivers in the upper part of the catchment area are usually clean but on their way towards the sea they frequently act as recipients and carriers for sewage and industrial effluents which cause deterioration in river quality. If there is only a small quantity of effluent, in relation to the total volume of natural flow, then pollution will be insignificant and the river will maintain its quality by the natural process of self-purification. Aeration and oxidation of the polluting organic material is an essential part of this process. Measurement of rates of aeration made in Britain and America has made it possible to produce mathematical models which take into account differences between rivers which are fast and shallow or slow and deep. It will be seen (Figure 5.5) that such modelling brings out the superior aeration in the fast, shallow river as compared with the slow, deep one.[7] Excessive discharges of effluent into a river will cause deterioration in quality and this may proceed progressively to the point where there are serious signs of pollution. These are manifest in loss of dissolved oxygen, increased turbidity, change in the flora and fauna culminating in the absence of fish, the occurrence of smell, and so forth.

In England and Wales a recent survey has distinguished four classes of river quality (Table 5.1).[8] This is not a classification of whole rivers but is rather one of

Table 5.1 *Comparison between biological classifications and chemical classifications of non-tidal rivers. England and Wales. (From HMSO 1972.)*

Biological class	Miles in chemical class				Total miles
	1	2	3	4	
A	10 241	284	20	0	10 545
B	2 107	1 228	99	7	3 441
C	452	554	427	183	1 616
D	27	129	202	572	930
Unclassified	4 173	1 095	323	190	5 781
Total miles	17 000	3 290	1 071	952	22 313

arbitrarily measured lengths of these, in the non-tidal stretches. Stretches of the rivers placed in Class 1 either receive no significant polluting discharges or, if they do, they are still well oxygenated and have a Biochemical Oxygen Demand of less than 3 mg/l. Class 2 stretches have substantially reduced oxygen content during the summer at times of low flow and are not in Class 1 on BOD grounds. They may receive turbid discharges but these have no great effect on the biology of the water. Class 3 stretches are of poor quality and have an even higher BOD; for considerable periods they have a dissolved oxygen saturation which is below 50 per cent. Finally, Class 4 stretches are considered to be grossly polluted and complete de-oxygenation may occur from time to time. Their BOD is 12 mg/l or more under average conditions.

These classes of river quality were made on chemical and biochemical criteria but a parallel biological assessment was also made. Stretches of rivers in biological class A have a diverse invertebrate fauna including stonefly nymphs and/or mayfly nymphs, caddis worms and *Gammarus*. Salmon, trout and grayling are present if local conditions favour them, that is, if the water is cool and there are suitable gravels for spawning, otherwise there is a good mixture of coarse fish. Class B stretches again have a good invertebrate fauna and coarse fish are present but trout, although they may be present, are rarely dominant. In class C stretches of rivers, mayfly nymphs and caddis worms are rare and although *Gammarus* can occur it tends to be replaced by the closely related form *Asellus aquaticus*. The fish are restricted mainly to roach and gudgeon. Finally, with a class D biological assessment, the invertebrate fauna is restricted to forms which are tolerant of pollution and low oxygenation, such as tubificid worms and chironomid larvae, especially *Chironomus thummi*. Fish are absent from class D stretches of rivers.

It will be seen from Table 5.1 that chemical class 1 and biological class A tend to go together, conversely chemical class 4 is often connected with biological class D, but there are river stretches where the chemical classification is more favourable than the biological one. For example, 27 miles are placed in category 1 D. The presence of such stretches will be explicable on the basis that there has been recent recovery from severe pollution but the fauna has not yet had time to show a corresponding favourable alteration.

When a major river runs through several different countries and receives sewage and industrial wastes from each of them there are political difficulties in surveillance of its whole condition. The Rhine is such a river and although its designation as 'Europe's majestic sewer' may not be completely justified, there is no doubt that its present condition is deplorable and further deterioration seems inevitable. In the Lower Rhine, near the Dutch border, the minimum oxygen content recorded fell from 4 to 1·5 mg/l in the period 1965-1969, as compared with a steady 8 mg/l in the Upper Rhine. This fall is attributed to increasing waste inflow from both French and German sources. On the French side the Moselle and Saar river basins discharge totally untreated wastes having an oxygen demand equivalent to sewage from 7·6 million people, proportioned as equivalent to 5·4 m industrial and 2·2 domestic. On the German side major cities still discharging virtually untreated sewage are Bonn, Cologne and Duisberg. The water is so murky that the planktonic algae can make very little contribution to oxygenation; only 7 g of oxygen per square metre of surface per day, as compared with 32 g in the River Danube for example. On the main stretch of the River Rhine there are no fewer than nineteen major nuclear and conventional power stations and more are planned. Those that use and return water for cooling purposes raise the temperature of the river; it is estimated that if they all operated in this way the summer water temperature would rise to 35 °C. Because of the lower oxygen-holding capacity of water at higher temperatures (see Chapter 1), this would lead to a serious oxygen deficit, in fact to total de-oxygenation at times of low flow. Therefore international co-operation in enforcing the incorporation of cooling towers seems essential, but the issue is controversial, some authorities maintaining that passage of water through power station turbines increases the oxygen content and that heat speeds up degradation processes, so assisting in self-purification. Another complication along the Rhine is the increasing activity of the potash mines in Alsace. These extract potassium chloride for sale, but since there is no sale for the sodium chloride which is extracted with it, this is put into the Rhine. The chloride content of the river at the Dutch frontier has accordingly doubled in the last thirty years and the long-term biological effects are not yet known but are not likely to be beneficial. At the 'receiving end' of the river, although the Dutch have created the huge freshwater reservoir Ijselmeer from the Zuyder Zee, their water demands are increasing to such an extent that large amounts of Rhine water will have to be used for drinking purposes by the turn of the century. They are faced with the prospect of having to remove salt and sewage, phosphate and nitrate, and special techniques may have to be devised to clear the water of organic compounds from industrial wastes, pesticides, etc., which defy extraction in existing waterworks purification practices.

References

1. H.B.N. HYNES, 1970. *The Ecology of Running Waters.* Liverpool University Press.
2. H.B.N. HYNES, 1960. *The Biology of Polluted Waters.* Liverpool University Press.
3. B. DOUGLAS, 1958. *J. Ecol.* **46**, 295-322.
4. H.B.N. HYNES, 1974. *Limnol. Oceanogr.* **19**, 92-9.
5. D.W. SUTCLIFFE and T.R. CARRICK, 1973. *Freshwat.Biol.* **3**, 437-62.
6. H.B.N. HYNES, 1955. *J. Anim. Ecol.* **24**, 352-87.
7. B.A. SOUTHGATE, 1969. *Water: Pollution and Conservation,* Thunderbird Enterprises Ltd.
8. HMSO, 1972. *Report of a River Pollution Survey of England and Wales in 1970.*

Chapter 6
Decomposition Cycles and Nutrient Re-Circulation

Various different kinds of micro-organisms occur in fresh water and almost all of them have some role in decomposition cycles and in nutrient re-circulation. Although the transformations which are accomplished are possibly of greater interest than the micro-organisms themselves, discussions of this aspect of freshwater biology have a distinct air of unreality unless the latter are also considered. In order to introduce both micro-organisms and their transformations simultaneously, it is convenient first to examine the aquatic fungi, particularly *Saprolegnia*, from the point of view of the life-cycle, occurrence and activity in the environment. Fungi are heterotrophic, working on elaborated organic materials to derive their energy sources, and the vast majority are also aerobic:

$$C_6H_{12}O_6 + 6\ O_2 \longrightarrow 6\ CO_2 + 6\ H_2O + 689\ 800\ \text{calories}$$

Glucose   Carbon dioxide

In the second half of this Chapter the bacteria are considered: here we have a simpler body organization but on the other hand considerably more physiological versatility is exhibited. Heterotrophic, aerobic bacteria are present in fresh water but anaerobic, fermentative types of heterotrophs also occur widely, especially in the bottom muds of lakes and rivers. (Fermentative types are very unusual in the fungi.) In addition autotrophic bacteria also occur; these derive their energy entirely from the transformation of simple inorganic chemicals. They are discussed in the sections devoted to the nitrogen, sulphur and iron cycles.

The Role of the Fungi
Saprolegnia

Of the fungi which mediate decomposition in fresh water, members of the Saprolegniales and Pythiaceae are always significant in any water body. The Saprolegniales is a large order of genera and species, but *Saprolegnia* itself is found most commonly. Fungi placed in this genus have a coarse and largely aseptate mycelium bearing sporangia, gemmae and oogonia (Figure 6.1). The sporangia are terminal cylindrical-shaped cells which liberate motile zoospores at maturity (Plate 8). These zoospore are biflagellate and are strongly aerotactic. When placed in a beaker of water they accumulate at the surface meniscus. This feature is a culminating expression of the strongly aerobic nature of the growth of the whole fungus, and indeed of aquatic fungi in general. The actual motility and taxis of the zoospore may have significance not only in its search for fresh aerobic situations but also in its final selection of a suitable substratum for growth. Substrata in fresh water which are particularly favoured are dead animal materials of all kinds, particularly such highly proteinaceous residues as dead fish, but also decaying vegetation to a lesser extent. Although such substrata may seem ostensibly small in amount, the estimation of *Saprolegnia* zoospores in natural waters gives fairly high values, of the order of hundreds per litre in stream and river water. There are however fewer in lake and reservoir water, usually only one to ten per litre.

In a lake such as Windermere there is a marked disparity between zoospore estimations in the littoral waters, which may approach stream and river levels, and estimations in the open water of the lake centre. This disparity constitutes some evidence that the zoospores are produced locally at sites of activity. As is the case with

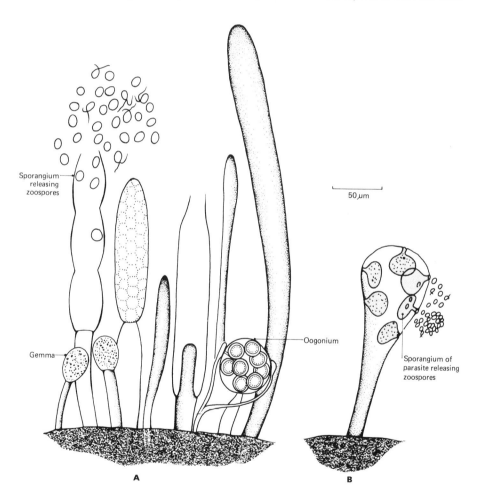

Figure 6.1 **A** *Saprolegnia* **B** sporangium of *Saprolegnia* attacked by parasitic fungus *Olipidiopsis*

many other aquatic micro-organisms, a wash-in effect from local soils can never be entirely discounted; *Saprolegnia* also occurs in soils. At Blelham Tarn where a large vertical cylinder of tough plastic is installed in the centre of the lake to prevent the entry of inflow stream water into this isolated water column, cumulative zoospore estimations are roughly four times as great in the open water as they are in the water of the tube. This implies that wash-in of *Saprolegnia* zoospores does occur into the lake via the inflows. In addition to wash-in from soil it is known that wash-in from the percolating filters of sewage treatment plants may also contribute *Saprolegnia* zoospores to the aquatic environment.[1] Apparently the fungus finds ideal conditions in this highly aerobic situation for its growth and sporulation. Presumably in this instance soluble nutrients are utilized rather than the more solid substrata favoured in natural environments. Treatment plants handling milk residues are especially noted for their *Saprolegnia* populations.

Dead fish exposed for even a few days in a reasonably unpolluted freshwater environment will demonstrate the growth of *Saprolegnia*, but it has been found that freshly collected water samples baited with split hemp (*Cannabis sativa*) seeds usually give good inoculations and dense mycelium bearing sporangia and oogonia is produced within a week at room temperature. These baiting recoveries on seeds demonstrate that nutrient-rich plant material is as available for colonization as is animal material. Seeds in particular have a high protein content, which is readily utilized.

The examination of dead fish illustrates the complexity of microbial cycles in nature. While *Saprolegnia* is the dominant heterotrophic colonizer, visible to the naked eye as a dirty white felt, it is accompanied by masses of bacteria and an associated micro-fauna. Many of the fungal cells autolyse within a short time under these conditions and other parasitic fungi such as *Olpidiopsis* may even move into the complex to attack the viable sporangial cells (Figure 6.1B). Using tellurite or antibiotics to eliminate bacteria, pure cultures of *Saprolegnia* are readily established on yeast-phosphate-soluble starch agar growth medium. Studies of pure cultures contribute further information.

Pure-culture characteristics of *Saprolegnia* fit into the pattern of comparative physiology established for the different orders of the aquatic Phycomyete fungi. Members of the Saprolegniales investigated so far, including *Saprolegnia* itself, can grow on a variety of carbohydrate sources. They utilize inorganic or organic nitrogen sources, require an organic sulphur source, but are fully autotrophic in their vitamin nutrition. The latter two characteristics are the special ones which distinguish Saprolegniales from, for example, the Chytridiales, another order of the aquatic Phycomycetes. When grown in pure culture, members of the Chytridiales can utilize inorganic sulphur, in the form of sulphate, for growth but often require a vitamin addition of thiamine (vitamin B_1) which they cannot synthesize themselves. In the present state of our knowledge it is impossible to relate these undoubtedly real differences between orders to ecological distributions or natural substrate preferences. There is no proper evidence for example that Saprolegniales have any competetive advantage in being able to inhabit substrates deficient in vitamins as compared with the Chytridiales which would seem to be at a disadvantage on the same material. There is always the possibility that heterotrophic fungi such as these may draw on dissolved substances in the water surrounding them rather than being completely dependent on the solid substrata to which they are attached. This possibility is heightened by recent findings on the nitrogen nutrition of aquatic fungi; this is discussed below.

The environmental conditions which can seemingly transform *Saprolegnia* from a harmless and indeed beneficial decomposer into an aggressive parasite are a matter of great current interest. It is true that the *Saprolegnia* which grows on living fish (see Chapter 4) has certain morphological characteristics which distinguish it from a number of purely saprophytic species, but the fact remains that fish parasitism is a purely facultative activity. *Saprolegnia* strains isolated from fish grow on synthetic media in the laboratory as readily as any others. Therefore it can be assumed that a saprophytic mode of existence is equally possible. When the living fish-*Saprolegnia* association is examined more closely it is known that the external mucus of fish, on which the fungus first becomes established, contains both carbohydrate and protein and even such free amino acids as alanine. These amino acids and the available sugars apparently enable the fungus to grow rapidly, and no doubt adaptive enzymes are produced later as the infection moves deeper into the body and the tissues of the latter are broken down. Natural protection for the fish may reside in the physical nature of the mucus, and its constant renewal, which generally seems to prevent lodgement of the *Saprolegnia* zoospore, together with the more doubtful existence of an effective antibody system. Evidence for the latter is seen in fish which make a natural recovery from infection, but this is rather rare.

Following exhaustion of the food materials in a natural substrate, fungi such as *Saprolegnia* move from a purely vegetative growth phase with the production of sporangia, into a resting phase. Resting phase structures distinguished in *Saprolegnia* are gemmae, which are merely walled-off sections of the mycelium, and oogonia. A sexual fertilization may be involved in the production of the latter (Figure 6.1) and the resultant oospores develop thick walls and are then highly resistant to desiccation and other adverse conditions. Resting structures constitute potential centres for further growth and spread, since they can germinate to produce sporangia and zoospores, but our knowledge of this aspect in the natural environment is almost negligible at present.

The decomposition of allochthonous materials by fungi

Materials which are imported into the aquatic environment from the terrestrial one may be very significant in freshwater energy budgets. Common imported, or allochthonous, materials are the leaves and wood of trees. In Lake Marion, Canada, very large amounts of such forest detritus enter via the inflow streams and calculation has shown that of a total lake energy budget of 280 g of carbon per square metre per annum, no less than 86 per cent is derived from the detritus, 10 per cent from benthic algae, and only 3 per cent from the phytoplankton and 1 per cent from aquatic plants. The limitation of the phytoplankton contribution to the energy budget is due to some extent to a peculiarity of Lake Marion imposed by its size and shape — this is the high rate of flushing through of the water. Any algal crops which develop are constantly diluted as a result. In European and North American streams and rivers, which are often tree-lined, the allochthonous plant contribution to the water is very obvious and recent work has suggested that the decomposing leaves may play a major part in sustaining the invertebrate fauna of such situations.

Having indicated that the biological turn-over of imported leaf and woody materials leads to enhanced productivity in fresh water, it is appropriate to examine the process in greater detail. When a tree leaf falls into water it loses its soluble sugars and amino acids rapidly, within a week in alder and oak, and outward leaching of organic acids also occurs. Ash leaves release 1·7 per cent of their weight as malic acid. The colonization of submerged leaves by aquatic fungi is very rapid and is complicated to a lesser extent than perhaps might have been expected by the prior occupation of the leaf surface by terrestrial fungi, while the leaf was still on the tree. On the tree, *Aureobasidium pullulans*, the black yeast, and *Epicoccum nigrum* are common components of the senescent leaf surface flora; once in the water these forms cannot compete with the newly arrived aquatic fungi.

Under normal conditions in streams and rivers of the British Isles, spores of members of the aquatic Hyphomycete group of fungi, e.g. *Clavariopsis aquatica* (Figure 6.2), are attached and growing on the submerged leaves within a week.

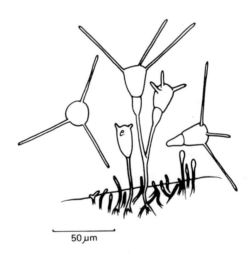

Figure 6.2 *Clavariopsis aquatica* showing spore release

These fungi seem to be particularly well adapted to locate their chosen substratum, having spores with long arms, often four in number, and arranged at a constant angle to provide a firm anchoring platform in any direction. They are able to attach efficiently even in turbulent waters and germination occurs through any of the arms. On wood some of the fungi, for example *Heliscus lugdunensis* (Figure 6.3), may grow so rapidly and build up such a compact pustule of growth, visible to the naked eye, that the conclusion is almost irresistible that these are acting as sugar

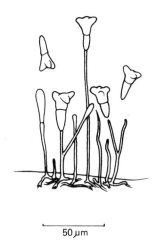

Figure 6.3 *Heliscus lugdunensis* showing spore release

fungi, i.e they have access to a very easily available energy source. Possibly the more long-standing nature of the leachable sugar reserves of wood, as compared with those in leaves, may be significant in this connexion. In the leaves, as was mentioned above, the most readily available sugars are leached out rapidly and the fungal colonists are compelled to produce a range of hydrolytic enzymes to gain and maintain a foot-hold; they never grow so luxuriantly as to be visible to the naked eye.

In addition to the aquatic Hyphomycetes, other fungi are early colonists on leaves, for example *Phytophthora* and *Pythium* species (Figures 6.4 and 6.5). In these fungi the spores (zoospores) have flagella and are motile, an adaptation to the aquatic environment which is believed to be advantageous in conferring a capacity to select or reject the final site for attachment and growth. In a study of the decay of alder, beech, elm, oak and willow leaves in the River Lune, Lancashire, it was shown that *Phytophthora* and *Pythium* species were present on all of these within a few weeks, and slightly later other motile-spored aquatic fungi, of the family Saprolegniaceae, also made their appearance. As decomposition proceeded, the outward appearance of the different leaves submerged in the River Lune changed to different extents. Starting from an initial immersion in the autumn, elm was extensively skeletonized in four to five months, while alder and willow were visibly eroded but oak remained ostensibly unchanged. Elm leaves became limp and slimy with the passage of time while the alder and oak leaves became increasingly brittle. These obvious differences in the pattern of decay, comparing one type of leaf with another, are correlated with feeding preference of the invertebrate animals which consume the leaves. This will be discussed again below; it is sufficient here to state that decaying elm leaves seem to be the most desirable leaf food.

Studies on the fungi in terrestrial environments have established the conception of succession, that is, of different fungi becoming predominant successively during the course of the decay of a particular material. Evidence for a fungal succession on leaves in water is not very great, even in oak leaves which may take two years or

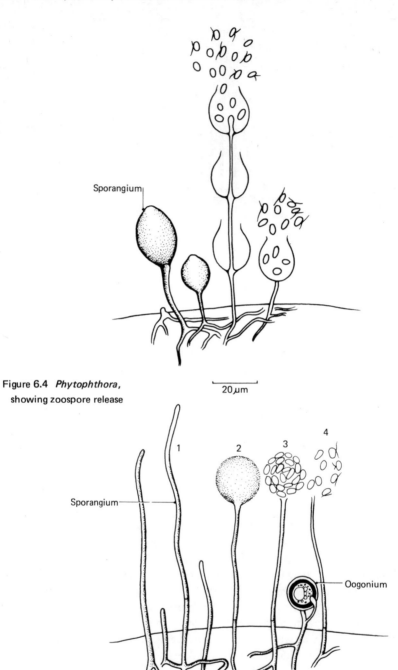

Figure 6.4 *Phytophthora*, showing zoospore release

Figure 6.5 *Pythium*, showing successive stages (1–4) in zoospore release

more to disappear. However, one species, *Dimorphospora foliicola* (Figure 6.6), seems to become very deep-seated and important with the passage of time, although it may still have been an early colonist. Decaying wood is often in the aquatic environment for a very long time before it disappears. In addition to the aquatic Hyphomycete colonists, often the same species as those which colonize leaves, other fungi become established later on the wood and eventually produce apothecia or perithecia, and fruit as Ascomycetes. There may be a life-history connexion between an aquatic Hyphomycete colonist and an Ascomycete colonist recorded later. Thus in a study at Smooth Beck, Lancashire it was shown that *Anguillospora longissima* was an important aquatic Hyphomycete colonist of alder, ash, oak and willow twigs — but under certain conditions it went on to produce ascospores in

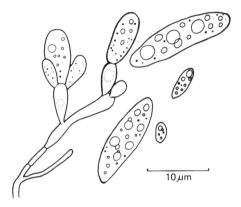

Figure 6.6 *Dimorphospora foliicola* (Note wide variation in size of released spores)

perithecia (perfect state) rather than conidiospores directly on the mycelium (Figure 6.7).

Since cellulose constitutes the main carbohydrate material in decaying leaves and wood in water, the expectation is that the associated fungi would all be shown to be cellulose-decomposers in pure culture. This holds for *Fusarium* but is not very obvious in most other cases. Here perhaps we have the difficulty of reproducing in vitro the exact background conditions which each particular fungus may require before its cellulase is evoked. Another fact of observation is the surprisingly large amounts of nitrogen contained by the shed and leached leaves, ranging from 0·7 per cent to 2·1 per cent; the protein levels are also substantial, ranging from 4·0 per cent to 10·8 per cent, all values being on a dry weight basis. So it would appear that a number of the leaf fungi may actually subsist on the purely proteinaceous materials which are present, never needing to hydrolyse the highly refractory polymers such as cellulose or cutin to obtain energy sources. Under natural conditions however it seems that this protein pool in the leaf may actually be conserved and not consumed by the fungi as they grow; it may even be enlarged. Kaushik and Hynes[2] showed that in a Canadian river the nitrogen content of submerged elm

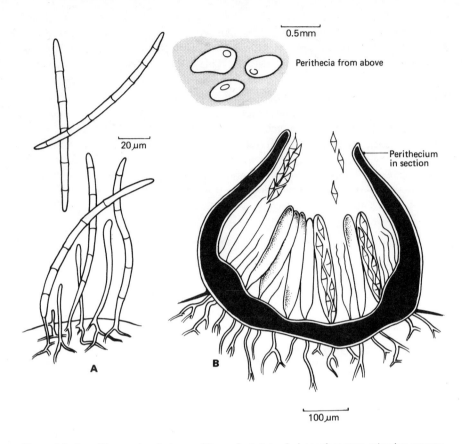

Figure 6.7 *Anguillospora longissima* and its perfect state **A** Imperfect state releasing septate spores **B** Perfect state releasing ascospores

leaves rose from 1·1 per cent to 2·4 per cent in four months. In an English Lake District stream, cellophane (pure cellulose) deposited as a 'model leaf' and carefully screened from silt deposition, increased the nitrogen content in a square of initial weight of 10 mg from 9×10^{-6} g N to 190×10^{-6} g N, in fifty days. This was an absolute as well as a percentage gain in nitrogen. The evidence that it is the fungi rather than the actinomycetes and bacteria which are mediating these nitrogen gains derives from two sources. Visual evidence demonstrates directly the presence of the fungi. In the Lake District stream experiment *Pythium* was the obvious colonist and was a presumptive cellulose-decomposer. In the Canadian experiments where river conditions were simulated in the laboratory, the addition of the anti-fungal antibiotics cycloheximide and nystatin reduced the nitrogen gains in the leaves while addition of the anti-bacterial antibiotics penicillin and streptomycin did not. It is presumed that soluble nitrogenous materials are extracted from the water and built into protein in the fungal cytoplasm, and here we see the inter-connected nature of freshwater microbiology; where the water is chemically more enriched so the likelihood of such nitrogen extraction is enhanced. Kaushik and Hynes have connected the observed preferences shown by *Gammarus* for different types of leaves as food with the differing nitrogen contents which the latter may contain. The animal prefers the leaf well decayed by fungi to an undecayed leaf, and gains more protein from it, comparing mouthful with mouthful (Figure 6.8). Leaves from different species of trees decay at different rates, and those decaying the fastest tend to be eaten first. Thus in the autumn, winter and spring, the main growing period for

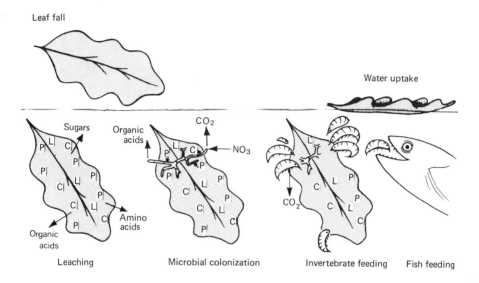

Figure 6.8 Transfers of carbon and nitrogen when an oak leaf falls into water. C, cellulose; L, lignin; P, protein

many stream animals in temperate climates, a steady supply of protein-rich food is available for them.

The decomposition of chitin by fungi and actinomycetes

Chitin is an important organic material which is produced and decayed in the freshwater environment. Its long-chain molecules are like those of cellulose in that a series of glucose units are bound together by β 1-4 linkages, but unlike those of cellulose in that an acetylated amino substituent containing nitrogen is attached to each glucose unit. Chitin is the chemical basis of the thickened shells which occur in many invertebrate groups of the Arthropoda, and it seems to be of special value in conferring rigidity on these protective exo-skeletons. In the freshwater Crustacea the chitinous exoskeletons are cast by numerous instar stages in addition to the adults, while in the aquatic Insecta the most spectacular shedding of chitinous exo-skeletons occurs when the pupa moults to the adult stage. In view of the large numbers of crustaceans and insects which inhabit freshwater situations, it will be appreciated that the total output of chitin is very large. Much of this is decayed and re-cycled but a fair proportion is not, and this becomes a part of the sedimentary deposits which will be discussed in a later chapter. Considering the chitin in the former category there is no doubt that cast exoskeletons, or exuviae, of insect pupae are particularly amenable to decomposition. On a calm summer day the undisturbed water surface of a lake may often be seen to be studded with floating insect exuviae, while in rougher weather they drift into the reed-beds. Some of the aquatic fungi which decompose the chitin in exuviae were named by the Danish mycologist, Henning Peterson, as long ago as 1904, and it was suggested then that these forms are so specialized in their physiology that they are probably confined to this peculiar environmental niche. These are representatives of the lower aquatic fungi of the order Chytridiales, the chytrids, with a very simple morphological organization, consisting essentially of a single sporangium only (Figure 6.9). Although we know much more about the chytrids now, i.e. that some chitin-degrading

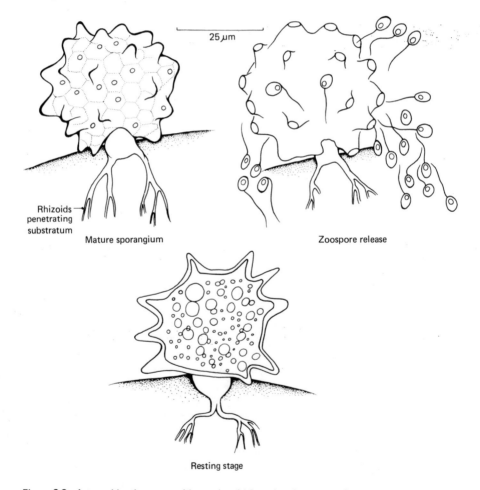

Figure 6.9 *Asterophlyctis sarcoptoides*, a chytrid found on insect exuviae

species have a more widespread activity, in the sediments as well as in floating exuviae, the fact remains that certain species have only ever been found in the latter materials.

In the laboratory study of chitin-degrading micro-organisms isolated from the freshwater environment, there are many aspects which are puzzling. Some of the fungi can utilize the chitin molecule, or its monomeric constituent glucosamine, as a nitrogen source but not as the sole carbon source to provide energy. In other words, although the amino substituent can be unlocked easily from the whole molecular chain, the latter cannot be easily disrupted by hydrolysis. However, if a high energy growth trigger such as yeast extract is supplied, then hydrolysis can proceed. It is difficult to see how such a mechanism could operate in nature where background nutrients are usually minimal. Actinomycetes which are isolated from the freshwater aquatic environment, and which are suspected to have a role in the decompositions there, often hydrolyse chitin. In fact this capacity is consistent enough to encourage the use of chitin agar as the standard isolation medium for this group of micro-organisms. Actinomycetes are a border-line group lying between the fungi and the bacteria; they resemble the former in producing mycelium, and sporangia in some species, but resemble the latter in that the cells are always very small and also in a number of physiological characters such as susceptibility to the same types of antibiotics which kill bacteria. Whether the laboratory findings of chitin utilization by actinomycetes can be deemed to have any ecological significance is not at all clear at present.

Representatives of the actinomycete genus *Actinoplanes* (Figure 6.10) are ex-

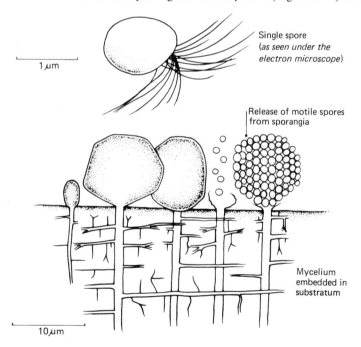

Figure 6.10 The aquatic actinomycete *Actinoplanes*

cellent chitin degraders, while at the same time the field experience is that *Actinoplanes* (Plate 8) has a positive predilection for decaying allochthonous vegetation as its growth substratum. Aquatic micro-organisms may sometimes have a degree of versatility in their exo-enzyme systems for which there is no obvious necessity. Chitin-degrading bacteria also occur in fresh water (see Plate 8B).

Fermentative fungi

In water, as on dry land, the vast majority of the fungi are aerobic and require oxygen for their growth. However, exceptions are always interesting and we now examine two aquatic fungi which are indifferent to the presence of oxygen; they exhibit a fermentative rather than an oxidative type of metabolism. In addition to this common propensity both show responses to the presence of carbon dioxide in the environment. The fungi in question are *Blastocladia pringsheimii* and *Aqualinderella fermentans*.

Blastocladia pringsheimii has been obtained from ponds all over the world, including Great Britain, generally by baiting with fruits such as apples or tomatoes. The baits are left submerged in wire cages for up to four weeks, and on recovery white macroscopic pustules are almost invariably observed. Although fruits do undoubtedly find their way into ponds, and may even constitute the sole natural growth substratum, the wide distribution of this fungus suggests the possibility of other materials being colonized as well. There is no information on this aspect so far. When grown in culture *Blastocladia* is shown to liberate lactic and succinic acids as the products of its glucose metabolism to gain energy, but no carbon dioxide is evolved. In nature both sporangia and resting spores are borne on a single thallus, the structure of which is somewhat reminiscent of a tree in that it has absorptive rhizoids simulating roots, and a stout central axis simulating a trunk. The sporangia and resting spores are borne on and between the branches which extend outwards and upwards from the central axis (Figure 6.11). When fruits which have been left deposited as baits are recovered, the *Blastocladia* pustules, consisting of large numbers of thalli massed together, are invariably clothed in swarms of bacteria and protozoa. When pure bacteria-free culture material yielded sporangia but no resting spores the effect of this bacterial investment in nature was re-considered. It was argued that bacterial metabolites, such as carbon dioxide, might constitute a constant factor in the environment and have some effect on the life-cycle of the fungus. Accordingly the fungus was placed in an atmosphere of almost pure carbon dioxide and the effect was dramatic. There was no inhibition of growth, such as would have occurred with the majority of fungi, and resting spores were produced in abundance. The effect was not duplicated when carbon dioxide was replaced by nitrogen and hence the former gas seems to have a definite role in resting spore initiation. When environmental conditions become particularly foul and anoxic in nature the fungus clearly responds by suspending vegetative growth and activity and producing resistant spores instead.

Aqualinderella fermentans also has a tree-like habit and has a more restricted distribution, not being reported from cold temperate regions such as the British

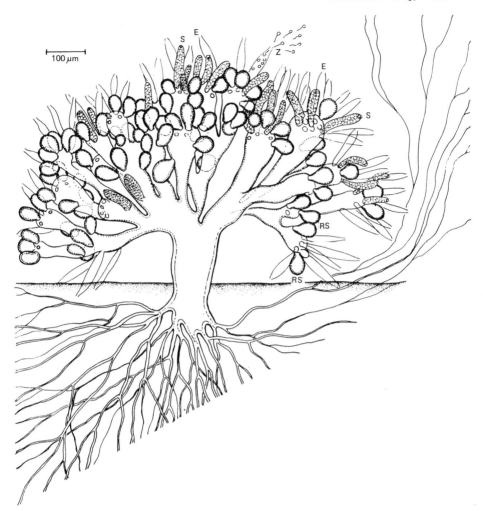

Figure 6.11 *Blastocladia pringsheimii* S, sporangium E, empty sporangium Z, zoospores RS, resting spore (drawn by Elizabeth M. Blackwell)

Isles, but occurring in tropical countries such as Costa Rica, Liberia and Nigeria. There are also reports of its occurrence from Louisiana in the southern USA. Once again recovery has been made by baiting with native fruits such as cashews and limes and there is greater evidence here that these do constitute the natural growth substratum, since they are locally abundant. *Aqualinderella fermentans* is similar to *Blastocladia pringsheimii* in that it produces a large tree-type thallus bearing the sporangia and resting spores. In culture *Aqualinderella fermentans* grows equally well in aerobic and anaerobic conditions, but only if a high level of carbon dioxide is provided. Like *Blastocladia* it produces only organic acids and no carbon dioxide from its glucose metabolism—its failure to produce its own carbon dioxide may explain its great avidity for the gas from the external environment.[3] In this connection it is suspected that most if not all heterotrophic cells are dependent on carbon dioxide fixation reactions which play important roles in a variety of metabolic transformations. In most of these cells, carbon dioxide fixation will be masked by an excess of carbon dioxide produced in aerobic respiration or fermentation. In *Aqualinderella* this requirement for carbon dioxide is more clearly exposed. How does its carbon dioxide requirement fit in with its ecological status? A recent study of its distribution in Nigeria suggested that it occurred predominantly in stagnant ponds where there was at least 26 mg/l of free carbon dioxide in the water but less than 4·8 mg/l of dissolved oxygen. Unpolluted lake water in equilibrium with the atmosphere at sea level has a free carbon dioxide content of only 0·6 mg/l but an oxygen content of at least 7 mg/l. There was also an apparently high requirement by *Aqualinderella* for dissolved organic matter in the water, at least 8 mg/l had to be present before the fungus could be expected to make its appearance. The evidence is that the fungus has evolved its particular metabolic pattern to suit specialized conditions. These evidently occur with sufficient frequency for the fungus not only to survive but even to compete successfully with the associated bacterial micro-flora. Examples discussed below will illustrate the metabolic versatility of the bacteria in fresh water; the fungi are generally more conservative in adhering to their heterotrophic and oxidative type of metabolism. However, *Aqualinderella fermentans* demonstrates that the fungi also can be versatile.

The Role of the Bacteria

Heterotrophic forms

The heterotrophic bacteria, which decompose organic material by aerobic processes, are important in fresh water. They are always present in association with the various fungi which have already been described, when the latter are breaking down the larger pieces of organic material such as dead fish or wood. Although we are reasonably confident that such substrata are the major ones for the fungi in fresh water, the smaller size of bacterial cells and their general lack of mycelial or rhizoidal systems makes it possible for them to participate in decompositions where the substrata are correspondingly smaller. A single decaying algal cell can nourish many bacterial cells. There is also the possibility that bacteria can work almost as free-living entities in the water, attached to the finest particulate matter or even entirely planktonic, utilizing dissolved materials such as those secreted by algae, or utilizing humic and fulvic acids from various sources. This conception of an active bacterial population freely suspended in the water colours much of the present work in freshwater bacteriology. A standard method of investigation involves plating small water samples or dilutions of these on an agar growth medium and examining and counting the resultant growth of colonies (Figure 6.12). The assumption is that each colony is derived from a single live bacterial cell, and colony counts therefore estimate the numbers of heterotrophic bacteria which were present in the water sample. A complication is that this viability count does not distinguish the heterotrophic bacteria which are actually active in fresh water from those which are essentially alien to the environment but which have recently arrived there by wash-in from the soils of the surrounding drainage basin, following heavy rainfall. The

Freshwater Biology 119

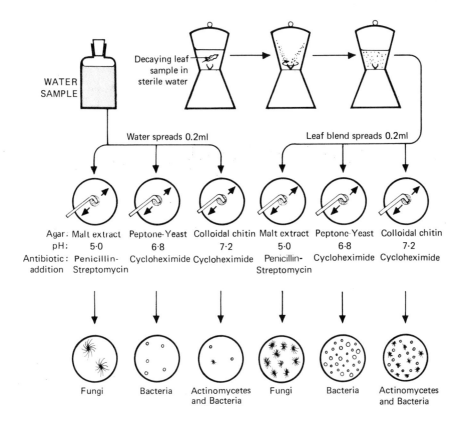

Colony yields after incubation at 20°C

Figure 6.12 A schematic examination of water and decaying vegetation for aerobic decomposers in the three categories: fungi, bacteria, actinomycetes. The different agars used, in conjunction with the antibiotic additions, act selectively, resulting in maximum yields for each category

latter bacteria are still viable for a time, and hence develop on the growth medium, but they may in fact have no role in decomposition processes in the water. Although these wash-in bacteria may therefore seem to be of little account in the ecology of a lake, they undoubtedly do have some significance since they continue to respire, even although not necessarily to grow, and thus deplete the dissolved oxygen store in the water. In a study at Blelham Tarn in the Lake District there was strong evidence that the wash-in of viable heterotrophic bacteria could be very large at all seasons of the year, but the major influxes always followed heavy rainfall or even the melting of snow and ice.[1] This wash-in occurred via the inflow streams. In the winter the washed-in heterotrophic bacteria were distributed throughout the water of the lake. They hastened the establishment of anaerobic conditions in the hypolimnion during the summer months, when this lower mass of water eventually became isolated physically from the surface layers. In late summer they appeared to enter the epilimnion only. The Blelham Tarn study was made at a site where the rainfall is very high and the peculiar configuration of the drainage basin leads to wide fluctuations in water levels as the lake fills and then partially empties. In other lakes the extent of the wash-in of heterotrophic bacteria may be much less.

Such a lake is Lake Vechten in the Netherlands. This is twelve metres deep but is artificial, having been excavated in 1938. There are no inflow or outflow streams and water enters only by direct seepage from the adjacent soil. The influx of heterotrophic soil bacteria into Lake Vechten is therefore less than the corresponding influx into Blelham Tarn, where the drainage area is a natural amphitheatre of terrain which includes agricultural land. The distribution of heterotrophic bacteria in Lake Vechten was followed through an annual cycle, particular attention being paid to the zones in the water where these bacteria attained numbers greater than 10^9 living cells per litre (Figure 6.13). Such zones occurred in the lower water of of the lake at ten to twelve metres depth, in the months of March and April, the time of the onset of stratification. At the same time a concentration of detritus was observed at these depths, attributed to this material emerging from the bottom mud. This detritus could well be the growth substratum for the bacteria. During the establishment of lake stratification the hypolimnion water became increasingly anaerobic, attributed to the metabolic activity of these bacteria and to the purely chemical oxygen demand of the mud. In the summer, with the stratification fully established, and particularly when the weather was calm, zones of heterotrophic bacteria of 10^9 cells per litre occurred at the thermocline level in the lake. This was explained on the basis that decaying material was sinking from the active plankton to this level, but remaining buoyed-up there for a while due to a density difference of the water to that in the hypolimnion below. Aerobic bacterial growth was presumably taking place on this decaying material before it slipped through into the anaerobic

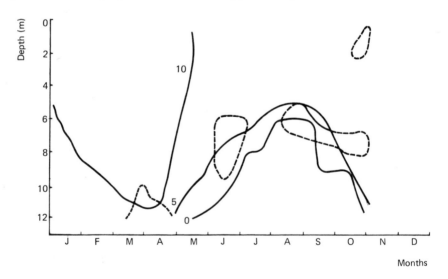

Figure 6.13 The seasonal distribution of heterotrophic bacteria in Lake Vechten. Areas enclosed within broken line denote bacterial concentrations of more than 10^9 bacteria per litre. The numbered lines represent oxygen concentrations in mg/l (after Cappenberg, 1972)

hypolimnion. Few heterotrophic bacteria in a viable conditon were recovered from the hypolimnion. After the autumnal overturn of the lake, zones of heterotrophic bacteria of 10^9 cells per litre were rarely encountered and there was a more general distribution throughout the water of the lake.[4]

At the thermocline of Lake Vechten in the summer season, the occurrence of a large concentration of aerobic heterotrophic bacteria, produced in response to phytoplankton decomposition, may well be paralleled by similar occurrences in other European lakes. However, because of the dynamic nature of the growth of plants and animals in the epilimnion water above, the balance between synthesis and decay in this particular region is worthy of special study. There is also a possibility that particular host-heterotrophic bacteria associations may occur in the epilimnion. In this connection the difficulty experienced in obtaining bacteria-free cultures of planktonic algae is noted; the suggestion has been made that if syntrophic partnerships between such living algae and bacteria do occur in nature, then this difficulty is more explicable.

As a beginning in the examination of these complex relationships in the epilimnion, a recent study on Esthwaite Water and Windermere in the Lake District has considered various aspects of exo-enzyme production by the heterotrophic bacteria.[5] Although a basal growth medium was used to detect these, it also contained special additions of starch, gluten or tributyrin respectively. Amylase-producing bacteria gave hydrolysed zones (detected by iodine) in the starch-peptone agar, protease-producing bacteria gave clear zones in the gluten-peptone agar and lipase-producing bacteria gave clear zones in the tributyrin-peptone agar. Simultaneously with the bacterial estimations, the three enzymes were also searched for in the lake water, using purely chemical methods. During the investigation the bacteria obtained were predominantly Gram negative types. Amylase producers included strains of *Acinetobacter, Serratia,* and *Xanthomonas*; lipase-producers included strains of *Bacillus, Pseudomonas* (Figure 6.14B) and *Xanthomonas;* protease-producers included strains of *Achromobacter, Acinetobacter, Bacillus, Corynebacterium, Enterobacter, Flavobacterium, Micrococcus* (Figure 6.14A), *Pseudomonas, Vibrio* (Figure 6.14E) and *Xanthomonas.* Considering the numbers of amylase-producing bacteria present during the productive summer months, they were at a maximum in May and a smaller peak occurred at the end of August (Figure 6.15). When these bacterial occurrences were compared with those of the algae present in the same water, it was seen that the May maximum followed, but did not coincide with, an algal maximum in April when the diatoms *Asterionella* and *Tabellaria* were predominant. The late August peak of bacteria coincided with the early phase of another algal maximum, this time featuring *Ceratium* and blue-green algae. When the enzyme estimations from the water were considered, the peak value for amylase was detected just after the corresponding bacterial maximum. It was not possible to obtain such a clear picture for the lipase and protease-producing bacteria; numbers of these and their enzymes produced tended to vary in a manner which could not be easily rationalized. However, on the basis of the results obtained, a tentative conclusion was made that the vigorous algal growth may have stimulated the growth of the associated bacteria early in the growing sea-

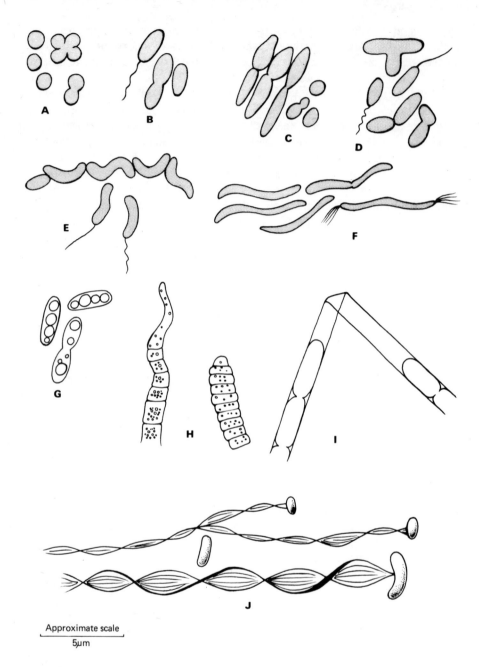

Figure 6.14 Freshwater bacteria A *Micrococcus* B *Pseudomonas* C *Azotobacter,* showing rod lozenge and spherical stages D *Nitrosomonas* E *Vibrio* F *Spirillum* G *Chromatium,* showing internal sulphur globules H *Beggiatoa,* two species both showing internal sulphur globules I *Sphaerotilus natans* showing cells inside transparent sheath J *Gallionella* showing cells at the ends of twisted ribbon-like excretion bands, which constitute holdfasts. The latter will later become encrusted with ferric hydroxide.

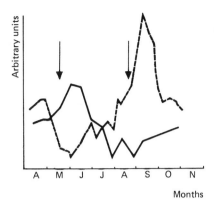

Figure 6.15 Surface water analyses at Esthwaite Water showing free amylase (line) and extracted chlorophyll (broken line). Arrows denote times of maxima from numerical determinations of amylase-producing bacteria (after Jones, 1971)

son but did so to a much lesser extent later in the growing season. Other factors of a purely physico-chemical nature may well be the stronger determinants of the growth of the heterotrophic bacteria in the epilimnion at this time.

The studies on heterotrophic bacteria described so far are analytical in that the population is not considered as a whole; the methodology involves plating water samples onto nutrient agars with the objective of isolating and growing each individual viable cell. A more integrated approach has been described recently.[6] Here the potential activity of all the heterotrophic micro-organisms in a water sample is examined by considering how an added substrate is utilized by the total population. Radioactive organic compounds, particularly glucose, are used in this work, glucose being almost universally utilized by heterotrophic micro-organisms. The fixation of radioactivity from glucose is measured by filtering out the micro-organisms and counting their activity at the end of the experiment: the $^{14}CO_2$ evolved in respiration is also collected and counted. These estimations of the rate at which the heterotrophic population can utilize glucose are of potential value as a sensitive indicator of their physiological state and activity at the time of collection of the water sample. The expectation is that bacteria which were alive but inactive at the time of collection remain so during the short incubation period used and most of the contribution to the carbon turn-over is made by the bacteria which were active initially. One river which has been monitored received sewage effluent and wastes from a phosphate mine but neither discharge created any detectable physical or chemical change in the river water and the bacterial counts showed no significant alteration downstream following the addition. However, the glucose utilization figures obtained from the samples taken along the river showed that there were peaks of uptake which could be related to the effluent additions. In comparing the potential value of this method in assessing bacterial activity as compared with the traditional method of plating on agar for bacteria, the gross and essentially statistical nature of the latter method is noted. Active and merely viable cells are not distinguished. The glucose utilization method aims at a high sensitivity; it may prove to be useful in comparative assessments of heterotrophic activity in natural waters.

The nitrogen cycle

Nitrogen is an essential constituent of all living matter, being a prominent component of the nucleic acids in all cell nuclei and of the proteins and amino acids which occur in the cell fluids. In fresh water the primary producers, the algae, utilize nitrogen in solution especially when it is in the form of nitrate. We have also noted above that aquatic fungi can apparently indulge in a similar activity. At the other end of the nitrogen cycle in fresh water, nitrogen compounds elaborated into protein are broken down by micro-organisms.

Any examination of the cycling of nitrogen through the freshwater environment is complicated by several factors.[7] Firstly there is always likely to be a considerable input from the drainage of adjacent land, and the effect is enhanced if the latter is fertilized. In some Lancashire rivers high nitrate levels in the water, up to 2 mg NO_3-N/1, are attributed to run-off from agricultural land. The input of nitrogen is relatively smaller from industrial and sewage effluents if these are present, contrasting to the situation for phosphorus where 90 per cent of the input from extraneous sources may be ascribed to them. Other complications in nitrogen study are that some aquatic micro-organisms can fix atmospheric nitrogen, thus increasing the store of this element in the environment; conversely, other micro-organisms which are present may liberate atmospheric nitrogen from nitrogenous materials. The latter process is known as denitrification.

A basic transformation involving nitrogen in the decomposition cycle is that of nitrogen mobilization, whereby organic matter, particularly its protein component, is broken down by micro-organisms. Both fungi and small animals (Protozoa and Crustacea) are involved in this, and nitrogen in the form of ammonium nitrogen (NH_4.N), free amino acids and other simple organic compounds is released by excretion or on the eventual death of the organism mediating the decomposition. Protein degradation by the fungi in fresh water was touched on in the account of *Saprolegnia* given earlier. Bacteria of the genera *Bacillus, Micrococcus* and *Pseudomonas* also play a considerable part in the activity and their transformations leading to NH_4.N release are often referred to as ammonification.

Nitrogen mobilization occurs both in aerobic and anaerobic conditions, the bacteria being primarily involved in the latter. Analyses suggest that in lakes which stratify, NH_4.N is at a maximum in the deeper waters during the summer but is at a maximum in the surface waters following the autumnal overturn.

Nitrification, the transformation of NH_4.N into nitrites and nitrates, is carried out by specific autotrophic bacteria which are only active in an aerobic environment, but which may survive in the absence of oxygen for long periods. *Nitrosomonas* (Figure 6.14D) converts ammonia into nitrite and *Nitrobacter* transforms the latter into nitrate. In the laboratory, *Nitrosomonas* and *Nitrobacter* have been notoriously difficult to cultivate due to certain features of their physiology including their high susceptibility to poisons such as metal ions, and even the amino acids in peptone may be toxic. Under entirely natural conditions it is suggested that in well-oxidized, stirred sediments, such as occur in the shallows of lakes and rivers, nitrification may contribute appreciable quantities of $NO_3.N$ to the water. Although it is difficult to produce evidence of this, laboratory experiments using vigorously stirred and aerated sediments have shown high nitrification rates, particularly if calcareous matter is present. In one experiment addition of 2-chloro-6 (trichloro-methyl) pyridine, a specific inhibitor for *Nitrosomonas*, stopped nitrification entirely and clearly pointed to this bacterium as the one initiating the transformation. Turning to conditions under human influence, nitrification is an important factor in sewage treatment and purification, since ammonia is such a toxic material. Its transformation in rivers is an essential part of the self-purification process.

Nitrogen fixation, the utilization of atmospheric nitrogen by organisms to build amino acids for growth, is carried out in fresh water by blue-green algae. This was discussed in Chapter 2. In addition, certain photosynthetic bacteria such as species of *Chromatium* can also fix nitrogen. *Azotobacter* (Figure 6.14C) is a heterotrophic bacterium, growing on decaying organic materials, which fixes atmospheric nitrogen under aerobic conditions. In the treatment of industrial wastes in sewage works the presence of *Azotobacter* might be very beneficial if the material to be degraded by biological treatment has a high carbon content but a low nitrogen content. Such a material is citrus fruit wastes. The net increase of nitrogen to the system made by *Azotobacter* could enable enhanced growth of micro-organisms to occur, resulting in enhanced degradation. To achieve this end, inoculations with *Azotobacter* have been attempted in pilot experiments, bearing in mind that good aeration is essential. Due to difficulties in analysis it is not yet known whether nitrogen fixation actually occurs in normal sewage treatment processes. Another type of bacterial fixation of nitrogen is mediated by *Clostridium.* Species of this genus are also heterotrophic but are anaerobic in addition; consequently Clostridia are only active in de-oxygenated waters, such as those that occur in stratified lakes in the summer season. They also occur more generally in freshwater sediments, always below the surface aerobic zone.

The key to denitrification is nitrate respiration. A variety of bacteria of the genera *Achromobacter, Bacillus, Denitrobacillus, Micrococcus, Pseudomonas* and *Spirillum* (Figure 6.14F) have developed the ability to use nitrate as an alternative to oxygen for respiration, and this results in the production of nitrogenous products with a reduced oxygen content. These products, the nature of which depend on the organism involved, are nitrite, ammonia, nitrous oxide or nitrogen. Recent work has established a strong homology between the aerobic (molecular O_2) respiration and the nitrate respiration of these organisms, and the same electron-transfer pathways are used in both until the very last steps. From the foregoing discussion it is suggested that denitrification is an anaerobic process, and observations show that this is so. The importance of denitrification in the natural freshwater environment is deduced rather than definitely proved at present. For example in Loch Leven, Scotland, up to 42 per cent of the input nitrogen cannot be accounted for subsequently, and this loss is at present ascribed to denitrification. There are situations where denitrification cycles in fresh water may be beneficial. In sewage treatment plants the aim is to achieve nitrification initially, but onward passage of nitrates in the effluent into a river may be almost as undesirable as that of ammonia. Excessive algal and weed growth will ensue, leading to oxygen depletion and other undesirable results. Accordingly, pilot scale experiments are in progress to examine the feasibility of encouraging a denitrifying micro-flora in the final stage effluent.

The $NO_3.N$ balance in a water body is often a delicate one and at any given time is governed by the relative rates of loss (utilization by growing plants and micro-organisms — denitrification) compared with the rate of generation (drainage water input, nitrification, nitrogen fixation). In the River Thames the balance of nitrification and denitrification was shown to be very important. Nitrate formed from ammonia in the highly aerobic parts of the estuary and the tributaries contributed 25 per cent of the oxygen used for oxidation of the organic matter in the anaerobic reaches.

The sulphur cycle

Bacteria playing a part in the sulphur cycle in fresh water are of three distinct types. First there are the forms which reduce sulphates to hydrogen sulphide in the bottom muds of lakes, ponds and rivers when conditions are anaerobic. Then there are the chemoautotrophic (chemosynthetic) sulphur bacteria which add oxygen to inorganic sulphur compounds and by so doing derive their energy to reduce carbon dioxide in the environment to cell material. Finally there are sulphur bacteria which are photosynthetic as well as chemoautotrophic, i.e. they are photoautotrophic; they require both light and sulphur compounds for the reduction of carbon dioxide. These latter sulphur bacteria are anaerobic.

Considering the purely chemoautotrophic forms, these are exemplified by *Achromatium* and *Beggiatoa* (Figure 6.14H). In lakes where the lower water, or hypolimnion, becomes depleted of oxygen during the summer, they occur on the surface mud just as this summer stagnation phase begins. At this time only a low oxygen concentration persists there, about 0·15 to 0·30 mg/l. With the lowering of the oxygen concentration, reducing rather than oxidizing conditions begin to occur first just below and then at the surface of the mud. Under these circumstances hydrogen sulphide appears and eventually the environmental background is optimal for the development of the sulphur bacteria, the hydrogen sulphide being oxidised, resulting in obvious sulphur globules being deposited within the cells. The oxidation may be carried on to sulphates by other bacteria and, as indicated by the following two equations

$$2H_2S + O_2 \longrightarrow 2S + 2H_2O$$

$$2S + 3O_2 + 2H_2O \longrightarrow 2H_2SO_4$$

The photosynthetic types of sulphur bacteria are exemplified by *Chromatium*.[8] *Chromatium* (Figure 6.14G) is a purple sulphur bacterium and is classified in the Thiorhodaceae, a family of bacteria which contain bacteriochlorophyll with a carotenoid background in their cells. These bacteria are strictly anaerobic as mentioned above. Their activity as photosynthetic organisms is best understood by comparing their photosynthesis equations (1 and 2) with that of ordinary green plants (3).

(1) $\quad 2H_2S + CO_2 \xrightarrow{\text{light}} 2S + H_2O + (CH_2O)$

(2) $\quad H_2S + 2H_2O + 2CO_2 \xrightarrow{\text{light}} H_2SO_4 + 2(CH_2O)$

(3) $\quad 2H_2O + CO_2 \xrightarrow{\text{light}} O_2 + H_2O + (CH_2O)$

In each of these three equations CH_2O denotes carbohydrate synthesis, leading to cell material elaboration. It will be seen that all three equations are essentially similar and involve a hydrogen-transfer mechanism which is photochemically operated. In this hydrogen-transfer mechanism, CO_2 is the hydrogen acceptor in every case and is reduced to $(CH_2O) + H_2O$, with the simultaneous oxidation of the hydrogen donor. The latter is hydrogen sulphide (H_2S) in the photosynthetic bacteria, and water (H_2O) in the ordinary green plants.

Chromatium occurs in lakes when the lower water, or hypolimnion, becomes completely depleted of oxygen during the summer. Under these conditions hydrogen sulphide is present in the surface mud and in the water immediately above it. Provided enough light is also available for its photosynthetic activity at these lower depths, *Chromatium* can then grow well. The adjustment of their position in relation to both hydrogen sulphide and light optima may lead to the establishment of zones or 'discs' of growth of the photosynthetic sulphur bacteria within the lake water profile. In the Lunzer Obersee, Austria, water samples coloured peach-red by the bacteria have been observed in such restricted dispositions.

To consider more closely the question of optimum light conditions for the sulphur bacteria of the Thiorhodaceae, it was shown as long ago as 1883 that they could photosynthesize in the near-infrared region of the spectrum, at wavelengths around 900 nm. This region of the spectrum is far from those regions where normal green plants photosynthesize through their chlorophylls (see Chapter 2). It therefore became an issue as to whether photosynthesis could proceed without the latter being present at all. The current view is that energy absorbed from the near-infrared illumination by the accessory pigments of the sulphur bacteria is transferred to chlorophyll, which is also present. The latter then functions as the sole participant in the ensuing photochemical reaction. By this means the sulphur bacteria have made possible an extension of the spectrum for effective photosynthesis, and as a consequence they can grow in nature beneath a layer of unicellular green algae.

Members of the Thiorhodaceae such as *Chromatium* are fully autotrophic, in that light, CO_2, reduced sulphur in the form of H_2S or sulphide, and simple inorganic ions obtained from the environment are their sole requirements for growth. However, laboratory studies have shown that they may grow much better if an organic acid substrate such as malate is provided. With a continued requirement for light and the reduced sulphur source they would then be considered as photoheterotrophic rather than photoautotrophic organisms. This finding opens the question of whether they are ever strictly autotrophic in the natural environment, bearing in mind that even in very clean water there is always a small amount of dissolved organic material present. Their degree of autotrophy may well be a varying characteristic, dependent on the local availability and suitability of such material.

The iron cycle

Micro-organisms involved in the transformation of metal ions occur in fresh water, for example, the iron bacteria which deposit ferric hydroxide around themselves. The starting material for the transformation is ferrous iron, and since in this state it is readily oxidizable by purely chemical means the iron bacteria generally occur where there is a lack of oxygen. High acidity also inhibits chemical oxidation and the occurrence of this condition favours the iron bacteria also. In addition to iron, manganese may also be oxidized by certain of these organisms. Iron bacteria are found widely in the bottom deposits of water bodies where high microbiological activity leads to oxygen depletion, for example, in stratified lakes in the summer season when ferrous sulphide occurs at the mud-water interface. When the lake overturns in the autumn these iron bacteria are moved up from the lake bottom and are found in the water of the lake for a while. Some authorities believe that the bacteria are fully autotrophic and can gain sufficient energy from the $Fe^{2+} \longrightarrow Fe^{3+}$ oxidation to fix gaseous CO_2 and utilize it as the sole carbon source. This may be true for *Gallionella* (Figure 6.14J), where the iron is deposited on the holdfast of the cell rather than on the latter directly. However, it has been estimated that to produce 0·5 g of cellular carbon from CO_2, 224 g of ferrous iron would be required. The figures illustrate the problem which these bacteria face in obtaining their energy on a purely autotrophic basis and explain the vast accumulation of $Fe(OH)_3$ which they produce at a microscopic level. *Thiobacillus ferrooxidans* is another autotrophic form, oxidizing iron at pH 3·0, at which value chemical oxidation of iron does not occur readily as stated above, and finding amenable conditions in acid mine waters. In this species the iron is deposited directly on the cells. Other iron bacteria can undoubtedly utilize organic matter which is present in the environment and hence these forms are not fully autotrophic. For example *Sphaerotilus natans* (Figure 6.14I) will grow in culture when ferrous iron is present or when nutrient materials such as peptone or yeast extract are supplied. This bacterium is rod-shaped and may either exist naked in chains or in an iron-encrusted sheath. In the latter state it is often referred to as *Leptothrix ochracea*, illustrating the confusion caused by the variable nature of these organisms. This confusion is still not fully resolved. The role of *Sphaerotilus natans* in polluted environments will be discussed in Chapter 9. The presence of iron surrounding iron bacteria is demonstrated by the application of a solution of oxalic acid and potassium ferrocyanide; this gives a blue colouration as the positive result.

The methane oxidizers

Simultaneously with the study of the heterotrophic bacteria in Lake Vechten discussed above, the distribution of the methane oxidizing bacteria was also examined. In contrast to the heterotrophic bacteria which are potentially capable of decaying any moribund solid organic matter which happens to be available and may even, it is suspected, utilize purely dissolved organic materials present in the water, the methane oxidizing bacteria are a very specialized group. Methane is formed as an end product of fermentation by bacteria in anaerobic situations in lake muds. When the lake water is fully aerobic in the winter months the methane is only formed deep in the mud, but it may rise to the surface as bubbles if its tension becomes greater than the hydrostatic pressure of the water overlying it. It was found in Lake Vechten that methane dissolved in the water was only detectable when the lake was fully stratified in the summer months, and it was then present below the thermocline. At this time the methane oxidizing bacteria could survive and multiply up to concentrations of 10^5 cells per litre in the hypolimnion, utilizing dissolved methane levels of up to 10 ml per litre of water. This methane oxidation by the bacteria proceeded in the water in an oxygen tension which was very low, so low that it could not sustain the aerobic heterotrophic bacteria. Eventually even this low oxygen content was used up, particularly in the very lowest depths, and the methane oxidizing bacteria were no longer found there. There was evidence however that some of the population could switch to an anaerobic metabolism, presumably mediated by a hydrogen transfer mechanism, but this aspect is not fully worked out at present. With the complete overturn and re-oxygenation of the lake in the autumn, concentrations of methane oxidizing bacteria were no longer found in the water. The annual cycle of distribution therefore gives a good example of the activity of a very special group of micro-organisms, adapting its site of growth to changing conditions in a lake.

Soluble Phosphorus Circulation

The amount of soluble phosphorus which is present in fresh water is of great significance in that it influences the final productivity, whether this is expressed in terms of yields of algae, higher plants or fish. In certain circumstances, for example in fish ponds, a high productivity is desirable and phosphorus fertilization is used to attain this end. However, where the water is in lakes and reservoirs and is destined for industrial or domestic use a high productivity is generally undesirable. The growth of algae in particular imparts undesirable turbidity and tainting to the water and expensive filtration and other treatments may be necessary to clean it. Although many algae can use a variety of organic phosphorus compounds it is generally accepted that free orthophosphate-phosphorus is the major and most available form of phosphorus for growth. Where the water is in a comparatively unproductive lake or reservoir its concentration is often very low. For example, in Lake Kinneret, the main water supply of Israel, there are only 1 or 2 μg/l orthophosphate-phosphorus in the water; in Windermere, a new supply for Manchester, 2 µg/l of orthophosphate-phosphorus is the mean figure from analyses made over an annual cycle. This compares with a total phosphorus concentration of 10 µg/l in Lake Kinneret[9] and 7 µg/l in Windermere. It has recently been discovered that aquatic microorganisms, including planktonic algae and the bacteria associated with them, produce phosphatase enzymes which can liberate inorganic phosphate from organic phosphate esters and make it available for their growth. In lakes such as Kinneret and Windermere the orthophosphate release by phosphatases may be the major source of readily available phosphorus for growth of the plankton complex. This orthophosphate release mediated by naturally produced phosphatases has been examined by experimental methods. Freshly collected water samples are shaken with chloroform in a closed jar and then allowed to settle. The chloroform separates as the lower fraction of a two phase system, carrying down with it all the living and dead algal and bacterial cells which were present in the sample. The living cells are killed by the treatment. However, estimations of the orthophosphate-phosphorus content of the water in the jar, carried out over several days, indicate that this rises steeply to a constant level. The effect is attributed to the phosphatases which were present in the water sample, unaffected by chloroforming, continuing to release orthophosphate from the water. If the water sample is heat sterilized the phosphatase enzymes are inactivated and no increase of orthophosphate occurs in the jar (Figure 6.16).[10] In addition to such indirect demonstrations and measurements of the effects of phosphatases, the enzymes can be assayed directly in the freshly collected water samples, by measuring their ability

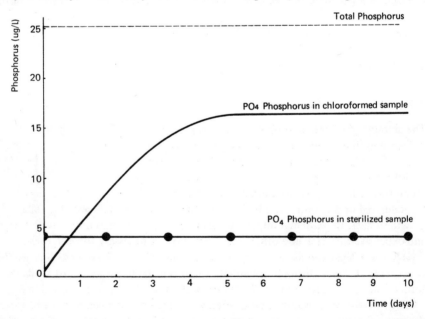

Figure 6.16 PO_4 release in a water sample following treatment with chloroform, mediated by naturally-occurring phosphatase (after Berman, 1969)

to catalyse the hydrolysis of p-nitrophenyl phosphate. It has been observed that the actual phosphatase-mediated release of orthophosphate in a water sample may be much less than that expected to be released by the amount of enzymes known to be present. The reasons for this discrepancy are under investigation because they may well be of crucial significance in the dynamic turn-over of phosphorus in the natural environment.

References

1. L. G. WILLOUGHBY and V. G. COLLINS, 1966. *Nova Hedwigia* **12**, 150-71.
2. N. K. KAUSHIK and N. B. N. HYNES, 1971. *Arch. Hydrobiol.* **68**, 465-515.
3. R. EMERSON and A. A. HELD, 1969. *Am. J. Bot.* **56**, 1103-20.
4. T. E. CAPPENBERG, 1972. *Hydrobiologia* **40**, 471-85.
5. J. G. JONES, 1971. *J. Ecol.* **59**, 593-613.
6. J. E. HOBBIE and C. C. CRAWFORD, 1969. *Verh. Internat. Verein. Limnol.* **17**, 725-30.
7. D. R. KEENEY, R. A. HERBERT and A. J. HOLDING, 1971. In: *Microbial Aspects of Pollution.* pp 181-200. Academic Press.
8. V. G. COLLINS, 1969. In: *Methods in Microbiology*, Volume 3 B, pp 1-52. Academic Press.
9. T. BERMAN and G. MOSES, 1972. *Hydrobiologia* **40**, 487-98.
10. T. BERMAN, 1969. *Nature, Lond.* **224**, 1231-2.

Chapter 7
Lakes, Reservoirs and Water Supplies

In Europe and North America there is a seasonal pattern in the variation of temperature and dissolved oxygen in lakes. This was discussed in Chapter 1 and mentioned again in Chapter 3. In Chapters 1 and 2 it was also suggested that lakes could differ in regard to their production of planktonic algae. The algal production, often termed the primary production, works down through the food webs to affect the growth of the animals. For example we saw in Chapter 3 how the *Chironomus anthracinus* larva, although living in the bottom mud, is sustained by the planktonic algae which fall to it from above. In their turn the *Chironomus* larvae consumed by fish contribute to a tertiary production level. The differing productivity of lakes is recognized in a classification scheme in which the range is from the unproductive, oligotrophic type to the highly productive, eutrophic type. Classical oligotrophic lakes are in upland situations, receiving water from barren drainage areas, for example Wastwater in The English Lake District. Eutrophic lakes generally have a more low-lying situation, are usually bordered by agricultural land, and may receive farmyard, human sewage or even industrial effluent. Blelham Tarn is an example of this type while Lake Esrom in Denmark lies somewhere between the two extremes.

Since a highly productive, eutrophic lake produces large phyto and zooplankton populations it is more likely to experience summer de-oxygenation in its lower depths than an oligotrophic lake of similar size. This is because much of this plankton eventually falls through to the hypolimnion and decays there, consuming oxygen in the process.

Reservoirs for the storage of water destined for domestic or industrial use may be perfectly natural lakes, or they may be of entirely artificial origin, from flooded valleys or even excavations with the water led into them (Plate 9). Lying between the natural and the artificial types are the lakes which have been dammed to increase their capacity – for example, Haweswater in the English Lake District.

Inevitably, because of their large size, reservoirs have the physical, chemical and biological characteristics of lakes but because the presence of plankton is an embarrassment from the point of view of producing a clean water, high productivity is undesirable. Hence the ideal lake to use as a reservoir will be of the upland, oligotrophic type. However, the shortage of such lakes to use and the ever-increasing demand for clean fresh water occasions the utilization of reservoirs which tend to be eutrophic. For example the London reservoirs are fed from rivers which contain sufficient dissolved nutrients to support a sizeable plankton. This must be removed before the water passes to the consumer. Again, reservoirs of this type will have the anaerobic strata in their lower depths (as described for Blelham Tarn in Chapter 1) in the summer season. This lower water contains the products of decay, including sulphides, which contribute undesirable tastes and smell and hence it is not generally abstracted. Clearly if the reservoir could be thoroughly mixed and aerated at this time a greater volume of acceptable water would be available and feasibility experiments have been carried out using pumps (see Figure 1.1B,C) on page 11). However, although mixing and aeration can be achieved there is also some stimulation of the algal growth. It may be at an acceptable level in some reservoir situations but not in others, and in the present state of our knowledge it is impossible to predict which pattern will emerge. This reservation about mixing has so far inhibited any wide-spread use of the system.

The Metropolitan Water Board supplying London has the advantage that it has several reservoirs at its disposal, and one can be withdrawn for 'cleaning' if biological productivity becomes to great. Copper salts are added to the water with the resultant death and sedimentation of the algae.

Water Treatment

In the treatment of water destined for drinking, cooking or other domestic purposes a major objective is to eliminate micro-organisms which are known to be dangerous to man as pathogens. However, there are other important aspects, for example, the removal of algae and the suppression of any unpleasant tastes or unacceptable odours. The most dangerous pathogens are the bacteria which cause cholera and typhoid fever and the viruses which cause poliomyelitis (infantile paralysis) and infective hepatitis. In a classic investigation in Hamburg in 1892, during a cholera epidemic, Koch showed that the city water supply, derived directly from the River Elbe, was contaminated with the cholera bacteria. In the contiguous town of Altona the water supply, from the same source, was sand-filtered before being used and no infection occurred. It is now recognized that the pathogens under discussion, both bacterial and viral, are present in the intestines of humans who are infected and are disseminated from there into sewage, and so may pass into a water supply. It therefore follows that any delicate indicator of the presence of human sewage in water will indicate the possible presence of these pathogens also. Conversely, it is argued that if a water supply is free of all trace of sewage then the pathogens also must be absent. In actual practice the sewage indicator used is a bacterium, *Escherichia coli*, a regular inhabitant of the human intestine and a non-pathogen. In order to test water for its presence, samples are incorporated into a liquid growth medium (MacConkey's broth) which contains bile salts and lactose. If *E. coli* is present it grows and produces both acid and gas from the lactose, while the bile salts inhibit the growth of any other lactose-splitting bacteria which may be present in the sample but are not of intestinal origin. Acid production is demonstrated by a colour change in a pH indicator incorporated in the broth; gas production occurs into a small inverted test tube (Durham tube) dropped into the larger tube of broth.

Determinations of *E. coli* cells in water are always made on a quantitative basis and as water passes through the various stages of its treatment en route to the consumer the 'coliform count' generally falls. The significance of this fall is that it will run parallel to a fall in the numbers of any bacterial pathogens which may be present, as indicated above. Finally, the coliform count is expected to be no more than one cell per 100 ml of sample, the standard laid down by the World Health Organization. Many primary water supplies contain treated (and not fully sterilized) sewage or even raw sewage, which contains enormous numbers of coliform bacteria, but the situation varies greatly from place to place. The water supply for the city of Liverpool originates from the uplands of Wales and is extremely clean. On the other hand, water supplies taken from some of the lower reaches of rivers in southern England contain 10 per cent or more of sewage or other effluent. It has been calculated that a supply taken from the River Ouse contained 13 per cent of sewage effluent in 1965, expected to rise to 28 per cent in 1981.[1] How is the water purified and the coliform count reduced to an acceptable level?

The treatment begins in an unspectacular manner, merely by storing the water in the reservoirs for as long a period as possible. The London supply from the River Thames and River Lee is stored in reservoirs with a capacity of 29 000 million gallons, and on average the water remains in them for three months. During this time the coliform bacteria count is reduced by 97·5 per cent during the winter and even more, by 99·8 per cent, during the summer, when 'self-purification' is more rapid because of the higher temperature. Self-purification is simply a question of cells which are adapted for life in a human body failing to find the requisite conditions to grow in water and slowly dying. The fall-off does not apply to bacteria only; the viruses are not as immortal as their mythology suggests and they also decline. The growth of microscopic algae in reservoirs can never be entirely prevented, and microstraining is a new method, first developed in Britain, for removing algae from water destined for drinking and other domestic purposes. The water is strained through a fine woven mesh of stainless steel with apertures small enough to hold back all but the very smallest cells. A microstraining unit consists of a drum covered on its cylindrical surface with the mesh, mounted horizontally and rotated (Figure 7.1). Raw water from the reservoir passes into the drum and out through the mesh while water jets wash away debris which accumulates on the inner surface.

In many water undertakings the water drawn off from the reservoirs is treated with aluminium sulphate, which brings colloidal materials such as the brown humic substances out of solution to be removed as sludge. Here we may note in passing that nitrate is not removed by the common processes of water treatment as they are practised at present. With the increasing application of nitrogenous fertilizers to agricultural soils, the nitrate content of river water is showing a general rise and there is a human health hazard, particularly to bottle-fed babies, in the consumption of a water with nitrate content greater than 45 mg/1. This level has hardly ever been approached as yet but clearly this is a problem for the future. Following aluminium sulphate treatment the water is sand-filtered; in the case of the London supply the sand filtration is particularly efficient and the preliminary aluminium

Freshwater Biology   135

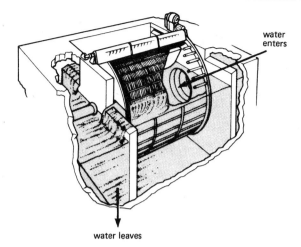

Figure 7.1 A microstraining unit for removing algae from a domestic water supply (reproduced by permission of F.W. Brackett & Co Ltd.)

sulphate treatment is omitted. Sand filtration is, as Koch observed, another barrier against bacterial pathogens and there is good evidence that viruses are impeded and removed in the process although more has still to be learned about this. Theoretically the viruses are so small that they should all pass through the pores in the sand beds; however it appears that forces of electrostatic attraction operate at the pore edges and trapping occurs as the viruses come near to them (Figure 7.2). Following reservoir storage and sand-filtration the water is expected to be of high quality and might be released for public use. However, sterilization

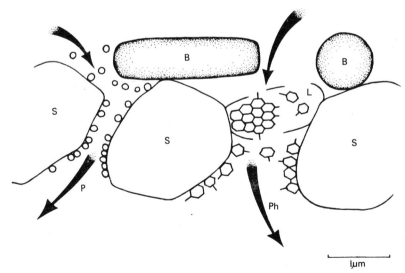

Figure 7.2 Hypothetical picture of sand filtration of bacteria and viruses: S, sand grain B, bacterial cell L, bacterial cell lysed by virus (bacteriophage) Ph, bacteriophage and P, poliomyelitis virus both trapped by electrostatic attraction (arrows indicate downward direction of water flow)

by chlorination is the final safety barrier against human pathogens and it is carried out almost universally.

Chlorination

The addition of chlorine or chlorine compounds to water to render it infection-free has been practised for a very long time; 'Eau de Javelles', a concoction of potassium hypochlorite, has been employed for this purpose in France since 1792. Although other chemical sterilants such as bromine or ozone have been tested with success and are even in limited use, most water supply undertakings continue to favour chlorination treatment in the present day. Chlorination treatment is cheap and has the advantage that its use over a long period of time has given a good understanding of its mode of action.

The passage of chlorine into water gives hydrochloric and hypochlorous acids

$$Cl_2 + H_2O \longrightarrow HCl + HOCl$$

and the latter may dissociate further

$$HOCl \rightleftharpoons OCl^- + H^+$$

The extent of this second reaction depends on the pH of the water and the significance of this is that HOCl is a more powerful disinfectant than OCl^-. Thus with a water pH of 4, HOCl predominates, while with a water pH of 10, OCl^- dominates the equilibrium. It follows that local knowledge of the water supply being treated is necessary in controlling the disinfection. Other complicating factors are the possible presence of organic matter and of ammonia in the water to be treated. Organic matter absorbs and destroys chlorine before it has time to exert a full germicidal effect on the water and it is therefore important that the sample is clear and bright before treatment commences. If ammonia is present then chlorine addition products such as chloramine are produced. These do act as disinfectants but less strongly than hypochlorous acid. Once again background knowledge of the water under treatment is required and there may even be a seasonal pattern for treatment, as the natural oxidation of ammonia in water held in storage reservoirs is less in the winter than in the summer. In some water undertakings the disinfection is made deliberately with chloramine, by successively passing ammonia and chlorine through the water, while in others the natural ammonia known to be present is utilized for this reaction. When chlorine is used itself as the disinfectant, the dose is generally 0·1 to 1·0 mg/l, depending on the quality of the water, but 5 mg/l or more may be required if there is much organic matter or ammonia present.[2]

1. B.A. SOUTHGATE, 1969. *Water: Pollution and Conservation.* Thunderbird Enterprises Ltd.
2. W.S. HOLDEN, 1970. *Water Treatment and Examination.* Longman Ltd.

Chapter 8
Sewage and Industrial Waste Disposal

In primitive societies, domestic sewage is re-cycled from man back to the land more or less directly but in civilized communities disposal by water has become traditional. In this system sewage is carried in all the water which the householder runs to waste and is piped to the treatment plant where it undergoes biological and chemical transformation before being passed as an effluent to a river, an estuary, or even directly to the sea. Although the piping of all sewage and domestic waste water into the sea would appear to be most satisfactory, there are good reasons, apart from economic considerations, why rivers should receive the effluents. For example, in southern England there are now so many wells and bore holes that the springs of a number of rivers have dried up – therefore water must be passed back into these rivers or they would be completely devoid of water, except in times of wet weather. The facts of the situation are that in England and Wales more than 1500 million gallons of sewage and waste water is produced per day and three quarters of this is discharged into inland waters. However, the greater part of this huge volume is of course pure water and the problem facing sanitary engineers is one of rendering the remaining material innocuous before releasing it as an effluent into fresh water. Sewage treatment as it is currently practised is itself an important facet of freshwater biology; furthermore, the biology of the river receiving the effluent is bound to be affected in some way, very profoundly if the treatment has been insufficient but less markedly if the treatment has been satisfactory.

The vital reactions, both biological and chemical, which need to be accomplished in the satisfactory treatment of sewage are essentially ones of oxidation. Because raw sewage contains enormous numbers of active bacteria, it exerts a high oxygen demand in respiration; it is essential that this demand be satisfied before the effluent is discharged into the river. If this cannot be done, severe de-oxygenation and death of most living things will occur in the river, together with all the undesirable phenomena associated with anaerobic conditions, for example, the release of hydrogen sulphide and other gases to the atmosphere in unpleasant smells.

The first stage in liquid sewage purification is sedimentation. Only 1 per cent of the initial volume is removed here but this comprises two thirds of the organic matter. Sedimentation yields a liquid sludge and this is dried, often in heated digesters at 30-35 °C for three or four weeks. This sludge will not pass forward for disposal in water and 'digestion' in anaerobic conditions is encouraged. Under these conditions 50 per cent of the organic matter is converted to gases, mainly carbon dioxide and methane, and the latter may actually be utilized to provide heating for the digesters and power for gas turbines for mechanical operations on the plant. The dried sludge has no obviously unpleasant characteristics and is sold or given to farmers as a fertilizer, heating having killed any potential pathogens, but final disposal in this way is not permissible if excessive amounts of heavy metals such as zinc are present. In this case the sludge must be disposed of in some other way.

Following sedimentation the liquid sewage moves on in the treatment plant to the activated sludge tanks, which are some 4 m deep and where the 'activation' is accomplished through forced aeration which keeps the whole mass in motion. The activated sludge process is a continuous one, constantly receiving fresh sewage and passing it forwards following aeration, but on an average any one portion of material has a residence of 4 to 8 h in the tank. During this time the bacteria which are present make what growth they can in the liquid, consume oxygen in so doing, and finally die. Protozoa are also active and consume enormous numbers of bacteria. Thus oxygen demand in the liquid falls as the bacterial mass declines; it is reduced still further because dead bacteria, protozoa and fine suspended matter which comes through the sedimentation process unchanged tends to aggregate or 'flocculate' into large masses which can then be removed and disposed of. In addition to the bacteria which show unspecialized growth on organic matter in the activated sludge tanks, others carry out vital transformations of nitrogen there. The main soluble nitrogen compounds in sewage are derived from urine and are ammonia and urea, the latter hydrolysing to ammonia. It is expected that the reaction

$$NH_4^+ \rightarrow NO_2^- \rightarrow NO_3^-$$

(see Chapter 6) will proceed in the activated sludge process and the efficiency of any particular treatment plant is often measured by the amount of transformation achieved in this direction. Since ammonia is a toxic compound to fish and other animals, it is important that it should not be released in large amounts in the final effluent.

Dissolved phosphorus is also an important component of sewage and it is inevitable that some of this passes through into the effluent. On the British scene this phosphorus is augmented by considerable amounts (often half the total) derived from detergents in domestic waste water which pass into the sewage works along with the sewage itself. With the future prospect that lowland river water, containing sewage effluent, will be used increasingly for water supplies (see Chapter 11), it becomes important to reduce this phosphorus to a minimum. If this is not done there will be massive algal growth in the reservoirs. Accordingly there has been considerable interest in activated sludge processes where operating conditions give a low phosphorus effluent. A famous instance is the Rilling Road plant at San Antonio, Texas, where particularly efficient extraction is achieved.[1] According to some authorities the aeration system is so efficient in this plant that the microorganisms indulge in 'luxury uptake' of phosphorus. But this is controversial and it is also possible that in the particular local conditions a change in dissolved carbon dioxide in the plant affects the pH and causes precipitation of finely suspended calcium phosphate.

Another method used to treat liquid sewage, an older one than that of activated sludge treatment, but still in widespread use, is the biological filtration method. Here the liquid passes downwards through a 'percolating filter' of clinker or stones (bundles of plastic tubes are beginning to replace these in some works) laid in a circular or rectangular bed two metres or so in depth. The sewage liquid is passed into the filter as a spray through rotating arms (circular bed) or arms which move backwards and forwards (rectangular bed). The rotating arm system is a familiar sight on railway journeys and a quick appraisal from a moving train can gauge the size of the sewage works from the number of beds being employed. A film of aerobic bacteria, protozoa and even small invertebrates builds up on the surfaces in the filter bed. Aerobic decomposition of the sewage liquid proceeds readily but difficulties may arise if the stone growth becomes too massive and sloughs off, blocking the stone interstices and so lowering the efficiency of the plant. The sloughed material may come through with the effluent as 'humus' and this is generally removed before onward passage is allowed to the recipient river.

In Britain, arrangements with the local health authority often result in wastes from industrial processes being passed into the sewage works. Substances such as phenols and effluents from textile manufacture and milk processing tend to degrade more efficiently if they are mixed with sewage, which gives a variety of subsidiary substances for the biological action of decomposition. Under these circumstances there is always the danger of overloading the sewage works, as industry expands, and considerable effort is expended in trying to prevent this. For example, in milk processing the dry collection of solid waste from the floors of the plant rather than hosing it down the drains will result in a reduced input to the sewage works.

Sewage works are never deliberately 'seeded' or inoculated with micro-organisms to break down the organic materials which they receive; they operate on the basic assumption that suitably active strains of micro-organisms will build up their own populations and if necessary mutate to strains giving activity against any new synthetic chemicals which may arrive. Generally this rather haphazard system is satisfactory. However, in 1949-1950, it became apparent that the new anionic detergents which had just been introduced into domestic use were passing through the treatment systems unchanged and causing obvious foam formation in the rivers receiving the effluents. Within the sewage works themselves the presence of the detergents prevented satisfactory oxidation in the activated sludge tanks. There was initial optimism that suitable decomposer strains would appear, but this hope was eventually abandoned. It seems that in these original 'hard' detergents the chief 'surface-active' agent was a sodium alkylbenzene sulphonate which included an alkyl chain of 12 carbon atoms, formed by polymerization of 4 molecules of propylene, $CH_3.CH:CH_2$. This alkyl chain had both straight chain linkages, for example, $\cdot CH_2.CH_2.CH\cdot_2$, and branched linkages, eg.

$$\cdot CH_2 - \underset{\underset{CH_3}{|}}{\overset{\overset{CH_3}{|}}{C}} - CH_2 \cdot$$

between the carbon atoms. Laboratory tests showed that the branched linkages could not be broken by micro-organisms, while the unbranched ones could. It was therefore assumed that if sulphonates with straight chain linkages only were used there was every likelihood that these would be attacked in the sewage works. Consultation with the manufacturers led to the introduction of such 'soft' detergents and by 1965 they had superseded the original ones. The desired result was obtained and river foaming largely disappeared; however, the situation is under constant review as other new materials come into use.[2]

As sewage passes through the treatment plant its degree of biological purification is measured by its Biochemical Oxygen Demand, or BOD. Raw sewage has a BOD of 600, that is to say that one litre of it consumes 600 mg of oxygen in 5 days at 20°C. This high oxygen demand of sewage, made by the bacteria which it contains acting on the organic matter, constitutes the greatest threat to river life if it is discharged directly. Severe de-oxygenation may result in the river, with the death of most living organisms. So the aim in sewage treatment is to reduce the BOD to such an extent that the final effluent has a value of 20 or less. The test is conducted in closed stoppered bottles, dissolved oxygen determinations being made initially and at the end of the 5 day incubation period. In actual practice the samples are usually prepared as dilutions with the addition of well-aerated water, so that the oxygen is not totally consumed during incubation. Phosphates and nitrates are often added to the bottles so that the bacteria will not be deficient of these nutrients as they exert their full effect.

Where sewage comes into the treatment plant with industrial effluent, as is often the case, attention has to be paid to the possibility of toxic materials being present which temporarily inhibit the activity of the bacteria and so reduce the oxygen demand. Discretion is therefore necessary in assessing the results of the test. Experience has shown that some waste materials from industry have a small BOD in 5 days but a very large one over a longer period. For example, wood pulp and sawdust, which decompose slowly, actually have a BOD of more than a million for total degradation. BOD measurements in sewage treatment plants have pin-pointed the vital role which the protozoa have as they consume bacteria in activated sludge. In an experiment, activated sludge without protozoa had 143 million bacteria per ml and a BOD of 61, while a parallel tank with sludge containing protozoa had only 4 million bacteria per ml and a BOD of only 17. In a sewage treatment plant a close surveillance of the vital BOD measurement in the various processes aims to ensure that the final effluent is of such a high quality that the river receiving it has a BOD which is not greater than 4.

There are enormous problems in the satisfactory disposal of sewage wastes from large centres of population, and solutions are evolved in the light of experience and experiment. Since 1889 and until recent times, sewage from the city of London passed through a sedimentation system but it was otherwise largely untreated before being discharged to the tidal estuary of the River Thames. Settled solids from the sedimentation were dumped into the sea. However, as the human population increased and industry also made an increasing contribution to waste discharge, the condition of the estuary deteriorated. This deterioration was seen in a diminution of the dissolved oxygen content of the water and the chemical reduction of sulphate to sulphide, with release of the noxious gas hydrogen sulphide into the air above. In the years between 1920 and 1940, in periods of hot weather and drought, dissolved oxygen fell to zero for short times in some places and by 1947 the worst reaches had no oxygen at all from July to September. The deteriorating situation culminated in that experienced in 1959 when the lower thirty miles of the river was completely anaerobic. At this time the large Northern Outfall sewage works, situated eleven miles below London Bridge, only operated the activated sludge process on a small fraction of the sewage received. More recently the activated sludge processing has been increased and with an improvement in the quality of the effluent passed to the estuary the condition of the latter has improved greatly. Anaerobic conditions are rare there now. The investigations on the River Thames brought out the important role of nitrate in polluted waters.[3] When anaerobic conditions first occur, if nitrate is present it is reduced before sulphate and will thus hold off the evolution of hydrogen sulphide. Although the anaerobic condition is itself bad, with the elimination of most plant and animal life, the hydrogen sulphide condition is infinitely worse. The practical implication is that the extent of nitrification of sewage during its treatment and prior to its discharge will have an important bearing on the subsequent condition of the water receiving it.

References

1. R.W. BAYLEY, 1970. *Wat. Treat. Exam.* **19**, 294-319.
2. B.A. SOUTHGATE, 1969, *Water: Pollution and Conservation.* Thunderbird Enterprises Ltd.
3. HMSO, 1964. *Water Pollution Research Technical Paper 11.*

Chapter 9
Pollution Phenomena

In Chapter 8 there was a discussion of some of the various measures which are in use to prevent or reduce the input of polluting materials into rivers. This may have given the impression that the situation is now well in hand, but unfortunately we do not live in a perfect world, and in Chapter 5 we saw that river pollution is still a current problem. In this Chapter river pollution is considered further, introducing conceptions and models which are proving helpful in understanding the biological aspects of the phenomenon. The saprobity conception can be represented as a model, the validity of which is justified by the following reasoning. A river is considered to be grossly polluted if it contains such a large amount of decomposable organic matter that the oxygen store in the water is greatly depleted, as the microbial flora and fauna acts on it. The discharge of untreated sewage effluent at one point into a river can lead to such a condition, termed polysaprobic, but generally the situation will improve downstream as the organic complex washes away and it is broken down successively to amino acids (α-mesosaprobic condition) and eventually to mineral salts (β-mesosaprobic condition). Finally, with full oxygenation completely restored the river is said to be in an oligosaprobic condition. The relationship of the resident flora and fauna to the saprobic condition is illustrated in the saprobiological triangle (Figure 9.1).[1] In the most intensely polysaprobic condition only anaerobic bacteria can grow, while at the other extreme, in the oligosaprobic condition, algae and higher plants are dominant. Within the triangle the height occupied by a particular category of organisms is proportional to numbers. Thus in the α-mesosaprobic condition there are fewer algae and higher plants than there are in the β-mesosaprobic condition. Pollution biologists recognize several other categories of saprobity in addition to those mentioned, and in the area of the triangle where the ciliates and flagellates dominate there are useful biological indicators of these different conditions. For example, there is a whole range of different *Vorticella* species with environmental preferences ranging through the complete spectrum from oligosaprobic to polysaprobic. With the international recognition of the various saprobic categories and the organisms which they contain, it becomes possible to initiate scientific dialogue. Workers from different countries can compare their findings and discuss the feasibility of remedial measures, if these are necessary.

A more comprehensive representation of the effects of the receipt of organic

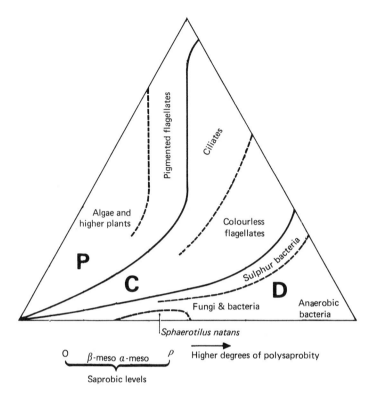

Figure 9.1 The saprobiological triangle: **P** producers **C** consumers **D** decomposers. The areas of the portions delimited are proportional to the number of organisms present (after Sládeček, 1972)

effluent is made in Figure 9.2.[2] Note the high Biochemical Oxygen Demand (BOD) of a poor quality effluent, falling away gradually downstream, and the oxidation of ammonia; this gives a corresponding downstream rise in the nitrate level, which may in its turn enhance the growth of algae.

Sewage fungus is the name given to the massive dirty-white complex of micro-organisms, visible to the naked eye, which often develops at or below the point of inflow of the organic pollutant. The latter may be treated or untreated domestic sewage or more commonly the residues from the paper or food and drink industries. Sewage fungus is in fact a mixture of micro-organisms, the proportions of which vary from place to place depending on the type of effluent present. For example, wood-pulping processing to make paper, using sulphites, may yield sulphides to the water, and in these circumstances sulphur bacteria such as *Beggiatoa* are conspicuous in the sewage fungus complex. The name sewage fungus is actually something of a misnomer because, although fungi (especially *Fusarium aqueductum* and *Geotrichum candidum*) may be part of it, the basic constituent is *Sphaerotilus natans*, a filamentous bacterium. In addition to this, other bacteria, especially gelatinous masses of *Zoogloea*, are usually present, as are also ciliated protozoa, of

144 Pollution Phenomena

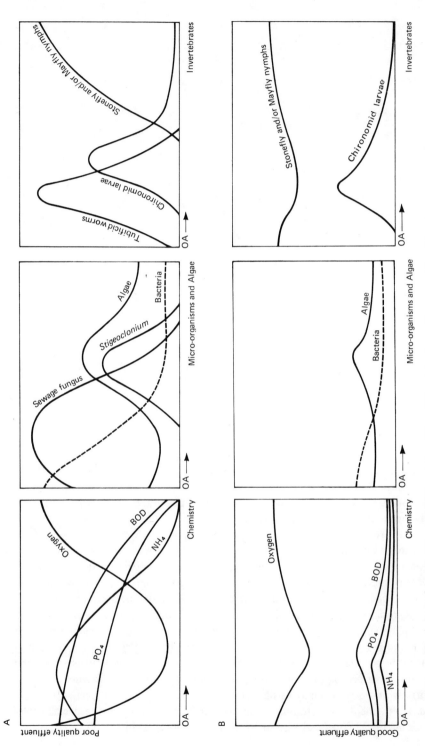

Figure 9.2 The effects of organic effluent additions, at positions OA, on rivers. **A** poor quality effluent **B** good quality effluent (arrows indicate direction downstream). (Based on Hughes, 1960.)

which the attached form *Carchesium polypinum* is found most frequently. A further element in the sewage fungus complex is often the green alga *Stigeoclonium* which seems to tolerate the other micro-organisms which clothe it in this situation for the sake of the additional nutrients which are available there (Plate 1). The *Sphaerotilus natans* constituent should be considered more closely — this bacterium is an attached form with a large expanded surface, ideally suited for the continual extraction of nutrients from dilute solutions, if conditions in the water are not excessively turbulent. In such conditions *S. natans* never becomes established while in waters with intermittent high velocities, as in river spates, the growth gets torn off and carried downstream. *S. natans* is an aerobic bacterium but it can grow with as little oxygen as 2 mg/l in a flowing system; thus it can tolerate river conditions where water oxygenation may be well below saturation levels (see Figure 9.2). This tolerance undoubtedly largely explains the success of *S. natans* in the sewage fungus complex.[3] The presence of sewage fungus in a river is undesirable from an aesthetic point of view and fish find it repellent, abandoning the area for cleaner water elsewhere.

Experiments in the USA have shown that if effluents causing sewage fungus growth are discharged into a river intermittently, then growth in the river is greatly reduced, the maximum period of discharge per day which must not be exceeded being six hours. However, the necessity for large storage facilities have so far limited the use of this method of disposal.

On the domestic scale a sewage fungus complex can be examined in material from sink outflows and drain covers. Here the material often lacks *S. natans* and true fungus (e.g. *Phoma*) binds a matrix of bacteria, protozoa and algae.

If the de-oxygenation is total, no normal river animals can survive although the

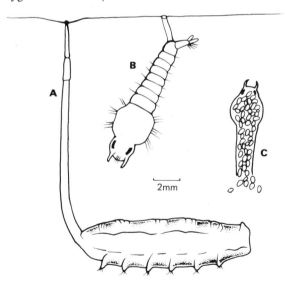

Figure 9.3 Insect larvae of **A** *Eristalis* and **B** mosquito, breathing by air tubes at the water surface. **C** mosquito has been attacked and killed by the fungus *Coelomomyces,* the resting spores of which occupy all the body cavities

larva (the rat-tailed maggot) of the insect *Eristalis* and the larva of the mosquito (*Culex*) may find a niche (Figure 9.3). These animals breathe through air tubes which make contact with the surface meniscus of the water. They also occur in foul ponds and manure-heap pools and this brings out the point that there is no special 'pollution fauna' of recent evolution. The animals which do occur in conditions made polluted by man's activity have had an apprenticeship in similar, but more natural, conditions elsewhere. Returning to Figure 9.3, we see that the tubificid worms and chironomid larvae become the dominant animals in the bottom mud when the effect of the pollution is most severe and in this connection we note especially their capacity to endure low oxygen levels. The release of a sewage effluent which has been very thoroughly treated will give a chironomid zone immediately, and sewage fungus and tubificid reaches are not distinguished (Figure 9.2).

Pesticides

With the widespread use of organic chemicals as herbicides, and insecticides in agriculture and industry, it is inevitable that some of these should find their way into fresh water. The most dangerous aspect of this is that certain of these materials do not decay and break down readily in the natural environment; once incorporated by a plant or animal in a food web they pass along this and accumulate in the highest member of the web. Here a lethal concentration may be attained, at a level several thousand times that found in the actual water. The compounds implicated most notoriously in such accumulations have been the organochlorines, (for example, aldrin, BHC, DDT and dieldrin), but legislation has now restricted their use. However, dieldrin is still used as a moth-proofing agent in the textile industry, etc. and can be detected in rivers in Britain. Toxicity tests on trout gave lethal levels only slightly higher than those known to be accumulated by the fish in nature, therefore in some areas there is a significant risk of fish mortality through this cause (Plate 10).

In the USA it has been noticed that when very large amounts of organochlorine pesticides are present in the water, most of the fish die but a small percentage are resistant, and this resistance, acquired through genetic selection, is preserved in subsequent generations. These resistant fish, which can carry large amounts of the pesticides, are lethal to the non-resistant piscivorous fish which consume them; they also constitute a dangerous health hazard if they are used as human food. In Britain, fish have had only low levels of exposure and it seems unlikely that resistant stocks have developed.

The organophosphorus pesticides which have largely come to replace the organochlorines are potentially less dangerous in nature because of their short persistence. However, little is known so far of their ultimate fate in fresh water and there is the possibility that their degradation products may themselves be toxic. This situation is under constant surveillance.

Herbicides provisionally approved for use against troublesome weeds in fresh water are dalapon, diquat, 2,4-D amine and maleic hydrazine; the latter seems particularly suitable since it is known not to accumulate through the food webs.[4]

Fish and River Pollution

It is recognized that fish are a very useful indicator of the state of purity of a river; no river is in a satisfactory condition unless fish live and thrive in it. On the other hand it has been said that nobody has ever seen a fish die from pollution caused by a sewage effluent. This is because the mobility of the fish, in contrast to the general immobility of the invertebrates and plants, enables them to move out of an unsatisfactory area. Even when there has been a large kill through a single accidental 'spill' of an industrial effluent, which the fish cannot escape, a part of the population downstream will usually survive as the offending material becomes diluted through the addition of stream or seepage water. These fish will work their way upstream later, when conditions improve. Some of the toxic effects of chemicals have been the result of man's past industrial activity. For example, in Wales seepage from dumps left by lead or zinc mines caused toxicity to fish long after the mines ceased to operate, after thirty-five years in some instances. Here there was a seasonal pattern. Caged fish left in the rivers were only killed during flood periods when seepage was at a maximum; at times of low water the caged fish survived in the pure stream water which surrounded them.

In recent times much effort has been expended in defining the maximum amounts of industrial effluents which can be allowed into a river without harming the fish and experimental exposures are made in laboratory tests.[5] These tests have shown that the chemical nature of the water receiving the effluent is of great importance. For example, the presence of calcium in the water is protective against metal toxicity. With a water hardness (see Chapter 1) of 12 mg $CaCO_3$/l, copper and zinc are toxic at the low levels of 0·04 and 0·4 mg/l respectively. With a water hardness of 320 mg $CaCO_3$/l, copper and zinc are only toxic at the much higher levels of 0·4 and 4·0 mg/l respectively. In other words, copper and zinc may be lethal to fish at very low levels if the water is soft. Conversely, ammonia is more toxic in hard waters than in soft ones, a difference connected with the differing degree of ionization of the molecule, and the physiological effect, in the two situations.

Laboratory experiments have also shown that the minimum amount of oxygen which a fish needs to survive rises as the dissolved carbon dioxide level rises. In situations of organic pollution where intense microbiological activity is taking place the enhanced carbon dioxide production may be the true barrier to the presence of fish rather than the low oxygen level itself.

Suspended solids are agents of biological pollution which tend to be underrated in their significance because they are often chemically inert. An example is the fine silt disturbed from a lake bed when sand is removed for industrial use. Underwater photosynthesis is inhibited and spawning sites for coarse fish may be rendered unsuitable. In a river the presence of suspended solids in its most extreme condition reduces the invertebrate fauna to tubificid worms and chironomid larvae only, and under these conditions gravel beds used for spawning by salmon and trout will have their interstices blocked, thus preventing essential aeration and causing the eggs to die.

References

1. V. SLÁDEČEK, 1972. *Verh. Internat. Verein. Limnol.* **18**, 896-902.
2. H.B.N. HYNES, 1960. *The Biology of Polluted Waters.* Liverpool University Press.
3. E.J.C. CURTIS, 1969. *Water Research* **3**, 289-311.
4. A.V. HOLDEN, 1972. *Proc. R. Soc. Lond. B.* **180**, 383-94.
5. HMSO, 1972. *Water Pollution Research, 1971.*

Chapter 10
The Sedimentary Record

As excavation on dry land can reveal evidence of past events, so too can historical information be obtained from probes deep into the underwater muds of lakes, estuaries and rivers. This historical information is interesting for its own sake but in addition it may cast light on trends of change in a water body which may allow prediction of events in the future. Such prediction could be of importance if, for example, the utilization of a lake as a reservoir is contemplated. Since lakes receive materials which are continually packing down and eventually being incorporated into their sedimentary deposits, they do in fact provide a potentially fruitful record of their own history. However, because of the difficulties of sampling and interpretation there are many apparent anomalies in the results obtained so far in this field of investigation. It is of crucial importance to examine sediment samples which are undisturbed in their vertical sequence and to this end special coring apparatus has been devised. The information obtained is derived from both chemical and biological analyses, particularly favoured subjects for the latter being diatom and pollen grain shells, both of which are highly resistant to decay and become incorporated into the sediments intact. Diatom shells have been particularly useful in following biological changes in lakes in very recent times (see below).

Sedimentation Interpretation

Some of the background to the chemical evidence of biological evolution in lakes will be discussed firstly, with particular reference to the English Lake District.[1] The lakes in this area were originally formed by glaciation of a dome-shaped terrain of Ordovician and Silurian rocks. Sediments deposited in them extend from the present mud surface to about 6 m below this in a fairly rich organic form (dark coloured sediments), but then show an abrupt change with further depth to a mainly inorganic form (light coloured sediments). This transition between the post-glacial organic deposits and glacial inorganic clays beneath them serves as a convenient base-line for dating. Radio-carbon and other supporting methods give this date as approximately 12 000 years before the present. Assuming that material has been deposited in the sediments at a constant rate over this whole period, which appears to be the case, it becomes possible to see the likelihood of deriving information about events which occurred in the lake during this time. In lakes in North-west Scotland which are exposed to exceptionally strong winds, the zone of wave erosion may extend down to 60 m and under these conditions much of the sediment which is deposited in an orderly sequence in the English Lake District lakes may never come to rest permanently and may be lost through the outflows. The sedimentary history of such lakes is accordingly very difficult to interpret.[2]

Although the total thickness of post-glacial deposits in lakes of the English Lake District varies slightly from lake to lake, this variation bears little relationship to the differing internal productivities of plant and animal life as they are known in the present day. For example, Ennerdale is much less productive than Esthwaite Water but its post-glacial deposit is thicker. It therefore seems likely that the thickness of the deposit is related to the rate of erosion of the drainage basin, which in turn is related to the nature of the water-shed. From this it is inferred that for an individual lake the major source of the sedimented material is the drainage basin of the lake rather than the material built up by photosynthesis and succeeding biological processes in its waters. This inference would seem to diminish interest in the deposits from the point of view of their being a historical record of events in the lake, but this is not the case. Meaningful information can be derived about the lake but it must always be considered against the background of material from the surrounding soils being constantly carried into it.

Chemical evidence of biological evolution

With regard to purely chemical events, the movement of manganese and iron in sediments is particularly interesting. In the sediment of Windermere south basin, the ratio of manganese to iron in the glacial clay is similar to that in igneous rocks, indicating a purely erosional transport of both elements. On passing upwards from the clay into the organic post-glacial deposits, however, a steep rise occurs in the manganese concentration to a level five times as high as that in igneous rocks. This enrichment of manganese, but not of iron, is interpreted as being the result of preferential removal of the former element from the developing soils of the drainage system with its subsequent deposition in the lake sediment. This preferential migration must have been brought about by the onset of reducing conditions in the soils which was of sufficient intensity to produce manganous iron but not of sufficient intensity to effect large-scale reduction of iron to ferrous ions. Manganese carried in solution into the lake basin, which presumably never became anaerobic, was oxidized there and eventually reached the sediment as manganese dioxide. This precipitation process is observed in the present-day since a black coating of manganese dioxide is seen on deep-water rocks and on any objects—for example, discarded pieces of crockery — which have lain undisturbed on the lake bed for ten years or more. In Esthwaite Water, a more productive lake at the present-day, and which regularly becomes anaerobic at the mud surface during the summer, the historical situation is initially similar to that in Windermere but then differs significantly. Thus the glacial clay again has an Mn : Fe ratio which is similar to that of igneous rocks, and on passing upwards into the organic deposits there is a steep rise in the manganese concentration. On passing upwards further, however, this rise is soon followed by an equally marked fall to a manganese content of only about a half of that occurring in the glacial clays. Thereafter the manganese content remains low right up to the sediment surface. It is deduced that the distribution of manganese in this sediment is related not only to the development of reducing soils in the drainage area but also to the development of periodically anaerobic conditions in the bottom of the lake itself. It is concluded that such conditions have probably occurred annually far back in the history of this lake, for about the last 8000 years in fact. During each summer since then manganese has been reduced to a soluble form and has migrated out of the lake with a resultant impoverishment of this element in the sediment. Present-day observations show clearly that manganese is in fact released from the sediments of Esthwaite Water during the summer period of anaerobiosis of the bottom waters and surface mud and is subsequently carried into the epilimnion at the overturn of the lake to be removed in the outflow stream (Figure 10.1). No such release of manganese occurs in Windermere south basin, which does not become anaerobic during the summer. Thus a purely chemical study of the sediments casts considerable light on the history of these two lakes; the evidence produced is that Esthwaite Water, but not Windermere, was a productive lake long before the activities of man could have been influential.

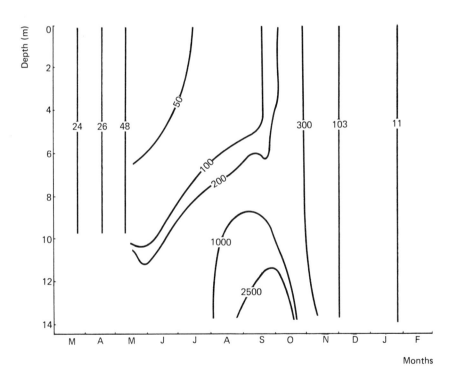

Figure 10.1 The concentration of manganese ($\mu g/l$) in Esthwaite Water through the year showing the release of this element into the lake as the result of summer anaerobiosis in the lower waters (simplified after Mackereth, 1966)

Plant Hydrocarbons

In addition to deductions on lake history which may be made from the inorganic elements present in deposits, other interpretations are derived from analyses for specific organic compounds. The saturated hydrocarbons, or alkanes, are particularly interesting. Methane (CH_4) is the simplest alkane and is a gas, but the higher members of this homologous series from C_{18} onwards are colourless solids and occur in plants. Higher plant residues yield major amounts of C_{27}, C_{29}, C_{31}, and C_{33} alkanes and only minor amounts of C_{26}, C_{28}, C_{30}, C_{32} and C_{34} alkanes. In most higher plants the maximum is at C_{31}. On the other hand the residues of lower terrestrial plants and freshwater algae yield alkanes in the C_{26}-C_{34} range with no preponderance of components with the odd number of carbon atoms. From such plants the C_{17} compound ($C_{17}H_{36}$, n-heptadecane) is often the main alkane component. Although plant material in the deposits may be decomposed and compacted to such an extent that it has no recognizable structure under the microscope, its presence may be inferred by the alkane analysis. Mud Lake in Florida, USA, is surrounded by semi-tropical vegetation. In sediments with an age of a few thousand years the dominant alkane distribution was shown to be in the C_{30} region with C_{27} and C_{29} showing maxima. This constituted evidence that higher plant residues had contributed largely to the sediment. On the other hand, an Eocene shale from the Green River, Colorado, USA, showed prominent C_{29} and C_{31} alkanes and C_{17} in addition. It was therefore possible to confirm the suspicion from other evidence that this shale had originally been laid down in shallow lakes.[3]

Plant Pigments

Chlorophylls and other pigment residues of plants can also be very persistent in sediments. Although aerobic decomposition processes in fresh water ultimately lead to the destruction of these materials, the processes apparently rarely go to completion before the pigments become inert in the lower anaerobic layers. An investigation in Canada began when leaves were still on the parent trees.[4] At this stage the identified pigments were seven in number and comprised β-carotene, chlorophyll-a and -b, lutein (xanthophyll), neoxanthin, phaeophytin-a and violaxanthin. Following autumnal changes, aspen and hazel lost their entire complement of pigment by the time of leaf fall but oak retained β-carotene, lutein and phaeophytin-a. This matched the observation that soil humus from oak forest contained these three latter pigments only. Again, in lake sediments, β-carotene and lutein were the only pigments to be found which were shown to be derived from leaves left deposited in perforated bags, while a much larger range of different pigments was shown to be derived from decaying algae and other aquatic organisms deposited. It was suggested that high ratios of β-carotene and lutein in comparison with other pigments recovered would indicate a large leaf input into a lake as compared with the contribution from lake organisms. Oligotrophic (unproductive) lakes might well show such ratios since the expectation is that the sedimentary matter in these would be dominated by imported materials such as leaves. Although this would be a useful pointer to the past history of a lake and its surrounding vegetation, there are several possible complicating factors. For example, green leaves blown into the lake by summer winds would possibly yield a larger and more varied pigment contribution to the sediments. Neoxanthin, derived from such an origin, was detected after a year of residence of leaves in a lake.

Observations at Esthwaite Water, in the English Lake District, have shown that pigments have been deposited and preserved in the sediments at a fairly constant rate over most of the last 12 000 years of post-glacial history. During an estimated last 150 years of deposition, however, the proportion of pigments to total organic matter has risen markedly and this is taken to reflect a more intense summer oxygen deficit at the lake mud surface during this time. Chlorophyll residues are no longer decayed at the rate they were previously.[5]

Animal Remains

Animal remains deposited in sediments have considerable potential as indicators of past lake history, although this potential is not fully exploited at present. However, crustacean and insect remains have already yielded provocative information. Of the Crustacea the Cladocera, especially *Bosmina*, and to a lesser extent *Daphnia* are well represented in the sediments but for reasons unkown remains of copepods such as *Cyclops* do not preserve. Since all these remains are essentially chitinous exoskeletons, microbial decomposition processes must act on them in different ways. However, bearing in mind the limitation which non-preservation imposes on the findings, the post-glacial deposits in Esthwaite Water have been extremely fruitful.[6] Within the 12 000 year time-span represented, no crustacean or insect exoskeleton has been recovered which cannot be identified with reference to a species living in the present day. However, the relative species composition in the deposits has altered during this time and this allows for speculation on past changes in the condition of the lake. For example, *Bosmina coregoni* appeared to be the dominant cladoceran in the plankton up to about 1000 years ago when it began to be replaced by *Bosmina longirostris,* although on the evidence of the deposits the two species can co-exist to some extent. Coincidentally with this change, the midge insect remains began to include substantial proportions of *Chironomus* in addition to *Tanytarsus-Sergentia*. These alterations in the balance of species are strongly suggestive of a major change in the biology of the lake at this time and it is suggested that the Norse immigrations and settlements, with the de-forestation which this entailed, were responsible. Presumably, chemical enrichment ensued from leaching from the soil and the upsurge of *Chironomus*, with its known powers of surviving oxygen deprivation in its larval stage (see Chapter 3), would perhaps suggest that only from then onwards was the lake hypolimnion anaerobic during every summer season. This continues to the present day although the influx of sewage is now contributive. The chemical evidence, as discussed previously, for the history of Esthwaite Water is that summer anaerobiosis had in fact been a feature of this lake from much earlier times, and this seems to conflict with the zoological evidence. Perhaps the precise degree and duration of anaerobiosis controls the benthic distribution of animals in such situations.

When one considers more recent changes in the invertebrate animal deposits in Esthwaite Water, the position of *Daphnia* is especially interesting. Only fragments of the body of *Daphnia* occur but the ephippia, the protective cases which enclose its sexually produced resting eggs, preserve well. It has been noticed that there was abundant deposition of ephippia during most of the history of the lake, but in more recent times, in the last hundred years or so, deposition has declined to vanishing point.[7] Well-fed *Daphnia* reproduces by parthenogenesis only, and in the present day in this lake this is in fact the sole method of reproduction. Therefore it has been argued that food for *Daphnia* is now more plentiful in the lake than it has ever been before. The dating of very recent deposits in lakes is discussed more fully in the next section.

Very Recent Deposits

Lake sediment deposits laid down in very recent historical times, in the last two hundred years or so, are of particular interest.[8] If such deposits could be dated with extreme accuracy and analysed for plant and animal remains, it might be possible to decide whether or not there has been a significant acceleration of biological change. The Mackereth short core sampler[9] has proved to be excellent for this work (Figure 10.2).

Because these sediments near the surface of Windermere are contaminated with ancient carbon derived from coal-burning in Britain, which has proceeded during this period but did not take place to any appreciable extent earlier in history, the ^{14}C dating method gives unacceptable figures. In another method of dating the recent sediment has been sectioned horizontally and examined for the intensity and direction of its magnetic field. The sections show an easterly or westerly deflection which must be a reflection of the polarity which iron took up during the time of its deposition in the lake. Since the easterly and westerly deflections occur in regular sequence they can be equated to the records of physicists and astronomers made independently in historical time. The last peak westerly deviation occurred at 1820 A.D. and this position is located 30 cm below the present mud surface. Another method of dating for the very recent sediments relies on the estimation of caesium-137. Since 1954, but not previously, Cs-137 has been present in the atmosphere and has been carried into the lake through rainfall. Since it does not move downwards into the lower sediments its position and amount can be correlated with atmospheric Cs-137 records obtained from other sources and dated accordingly.

The average rate of sediment deposition in Windermere over most of the postglacial period is only 0.5 mm per year. However, using the magnetic and caesium dating methods it would appear that since about 1850 A.D. the rate has accelerated tremendously, averaging 2 mm per year from this time onwards (Figure 10.3). At this dated level, approximately 25 cm below the present mud surface, the nature

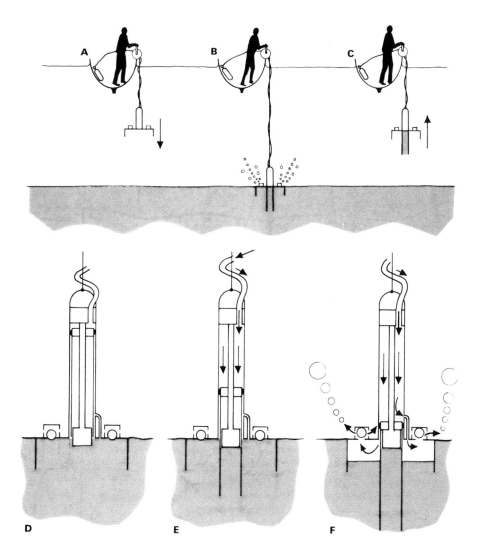

Figure 10.2 Operation of the Mackereth short core sampler for obtaining undisturbed sediment **A** the sampler descends to the lake bed with the coring tube withdrawn. **B** at the lake bed the tube extends into the sediment. **C** the sampler ascends with sample held in the tube by compaction. **D, E, F** show how compressed air extends the coring tube and then lifts the sampler above the mud

of the sediment also changes from brown to black. This change is thought to be a permanent one, not attributable to immaturity of the sediment. Again, at this level the silicon from diatom algae increases from 1.5 to 7.5 per cent of the sediment dry weight and recognizable *Asterionella* cells increase from 4 to 40 per cent of the total diatom count. It is possible to infer that this sudden change in the nature of the sediment, accompanied by an increased production of *Asterionella* in the lake,

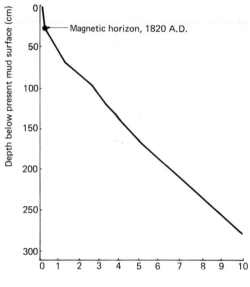

Figure 10.3 Dating of sediment in Windermere, South Basin, using ^{14}C at the lower depths and magnetic methods near the surface. (Note the steep rise in sediment accumulation since 1820 A.D.) (after Pennington, 1973)

was probably not due to the introduction of modern farm practices and the first additions of fertilizers at about 1800 A.D. It seems much more likely that since about 1850 the increased human population of the surrounding area and the influx of tourists has led to seepage of sewage into the watercourses and ultimately into the lake; this in its turn has led to a marked nutrient enrichment of the lake with profound, although subtle, changes in its ecology. Here we note that algal blooms of nuisance proportions do not yet occur in its waters but clearly the situation will require surveillance in the future if this lake is to be protected as a national water resource.

In lakes still surrounded by areas of little human habitation, for example Loch Clair and Loch Sionascaig in Scotland, there has been no acceleration of sediment accumulation in recent times.[2]

Summary

Sediments in lakes can provide chemical evidence of past events in their drainage basins and their own waters. Iron and manganese analyses suggest that one particular lake, Esthwaite Water, had an annual anaerobic phase in its deeper waters as long as 8000 years ago, before the activities of man could have been influential. Chemical analysis for organic compounds in sediments distinguishes several which can be related to plant growth and makes it possible to infer something of the biological history of the water body. Animal remains are a potential source of information, with the Crustacea and Insecta contributing most so far. The very recent sediment deposited in Windermere has been dated by special methods; analysis of the diatom shell content suggests that since about 1850 A.D. there has been an accelerating change in the ecology of this lake.

References

1. F.J.H. MACKERETH, 1966. *Phil. Trans. R. Soc. Ser. B* **250**, 165-213.
2. W. PENNINGTON, E.Y. HOWARTH, A.P. BONNY and J.P. LISHMAN, 1972, *Phil. Trans. R. Soc. Ser. B,* **264** 191-294.
3. R.B. JOHNS, T. BELSKY, E.D. McCARTHY, A.L. BURLINGAME, P. HAUG, H.K. SCHNOES, W. RICHTER and M. CALVIN, 1966. *Geochimica et Cosmochimica Acta* **30**, 1191-222.
4. J.E. SANGER and E. GORHAM, 1970. *Limnol. Oceanogr.* **15**, 59-69.
5. G.E. FOGG and J.H. BELCHER, 1961. *New Phytol.* **60**, 129-38.
6. C.E. GOULDEN, 1964. *Arch. Hydrobiol.* **60**, 1-52.
7. W.J.P. SMYLY, 1972. *Verh. Internat. Verein. Limnol.* **18**, 320-6.
8. W. PENNINGTON, 1973. *Freshwat. Biol.* **3**, 363-82.
9. F.J.H. MACKERETH, 1969. *Linnol. Oceanogr.* **14**, 145-51.

Chapter 11
The Way Ahead

In the future, studies in freshwater biology will continue to delight man's eye and provide him with intellectual nourishment. However, it is man's potential capacity to control and direct events in the freshwater environment which is the background for this Chapter. From the many aspects of this which could be discussed, three have been selected, illustrating some current work and its objectives in the fields of invertebrate zoology, fish farming and reservoir biology, respectively. From these discussions it will emerge that man's accumulated knowledge of freshwater life can be applied for a useful purpose. Our hope must be that such applications will continue to bear fruit — but not at the cost of the destruction of the natural environment — the priceless heritage of us all.

Mosquito Control

In the control of mosquitoes and the diseases they transmit, of which malaria ranks high in importance, the use of chemical insecticides will no doubt continue. However, the use of DDT has led to the evolution of mosquitoes which are resistant to it and has suggested that this and other chemicals may eventually become ineffective. Consequently there has been recent interest in the possibility of biological control, that is, the supervised attack by predators and parasites which occur naturally.[1] Clearly the aquatic larval stages of the mosquito must be the target ones in the life-cycle and extensive collecting has revealed that there are certain specialized fungi, of the genus *Coelomomyces,* which grow inside and cause their death (see Figure 9.3 on page 145). It is unfortunate that at the moment these fungal parasites cannot be grown outside their hosts in the laboratory; their preparation for use in the field involves the laboratory rearing of infected mosquito larvae. There are many gaps in our knowledge which bear on the cycle of infection, for example the route by which the fungal zoospore enters the larva is not known for certain and conditions suitable for infection can only be guessed at rather than defined at present. Here we begin to see the enormous difficulties which are inherent in all biological control work, but there is no doubt that increasing disenchantment with the more brutal chemical methods will lead to many other investigations of this type.

Fish Farming

An important recent development in fish farming is the rearing of Atlantic salmon on the west coast of Scotland.[2] This is a commercial investment and the salmon was chosen rather than any other species of fish because of its high price in the food market. In the final stages of this programme the fish are at present fed in a sea loch at Lochailort, where they are confined in large cubical pens with sides twenty feet long. This is the marine aspect of the operation but the freshwater one is equally important. This takes place at Invergarry where clean river water suitable for the development of the young fish is available. Selected mature fish from Lochailort are brought to the Invergarry hatchery and the fish are stripped to obtain fertilized ova. The successive stages of fry and parr are fed artficially and at the end of one year 30 per cent are silvered smolts and are ready for transfer to the sea pens. Experience has shown that if healthy smolts are selected this transfer can be made directly, without any acclimatization through increasing salt concentrations. Under the present conditions the remaining 70 per cent of the fertilized brood takes two years to develop into smolts at the hatchery, and because this is an economic operation research is directed to gain a higher percentage of one year smolts. Selective breeding, hormonal treatment and the use of warmed water are possibilities which come into the picture here. At present the output of mature salmon is quite small, 50 tons in 1973 and an expected 100 tons in 1974. However, with increasing expertise in the technology involved, and the conviction that the clear waters of the west Scottish coast offer a suitable arena for activity, it has been possible to set a target figure for annual production at 10 000 tons. Such an output would require the annual rearing of an estimated 12 million smolts; the operations in fresh water will clearly need considerable expansion to produce these.

Reservoir Biology

Increasing population and industry brings ever-increasing demand for clean fresh water and the problem of supplying future requirements in Britain is under active consideration. In this connection it has been proposed that lowland reservoirs should be created in estuarine sites. Possible sites are on the Dee estuary, at Morecambe Bay and on the Wash. They would have the advantage that normal river flow would not be affected. Although the main problems are ones of engineering technology there are also many biological aspects involving the prediction of future events.

The proposed Morecambe Bay barrage would enclose a large area of shallow fresh water, much of it only 5 m deep, and would draw from the Rivers Crake and Leven on the west side and the Rivers Kent and Beela on the east side (Figure 11.1).[3] An attempt has been made to forecast the amount and kind of plant and animal growth which would occur in the completed reservoir, bearing in mind the unexpected problems encountered in the Kariba reservoir, for example. In view of

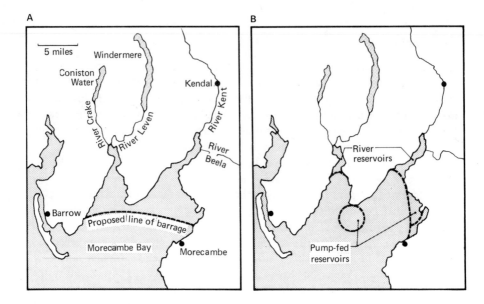

Figure 11.1 A Original plan for the Morecambe Bay barrage B modified scheme which has also been proposed (adapted from *Morecambe Bay: Estuary Storage,* HMSO, 1972)

the shallowness, the growth of filamentous benthic algae is possible, especially in the early stages when the brackish-water form *Enteromorpha* might be abundant for a while. However, since the planktonic algae pose the greater problem for water purification the potential growth of these was considered more fully.[4] To this end water samples from the main contributary rivers were incubated for 8 weeks at 10 °C, either as they stood or after inoculation with typical reservoir algae. Chlorophyll-a was then extracted from each sample and the results compared. It was apparent that the east side rivers, the Beela and particularly the Kent, had a far greater algal growth potential than the west side ones (Table 11.1). In explan-

Table 11.1 *Chlorophyll-a estimations (mg/l) in river water samples, after inoculation with algae and incubation. (From Corlett, 1972).*

River Leven	0.015
River Crake	0.025
River Beela	0.12
River Kent	0.25
River Thames	0.17

ation of this it was shown that water in the River Kent had a very high phosphorus content. Considered as a daily flow, the dissolved phosphorus was always the same, even although the amount of water in which it was contained varied greatly, depend-

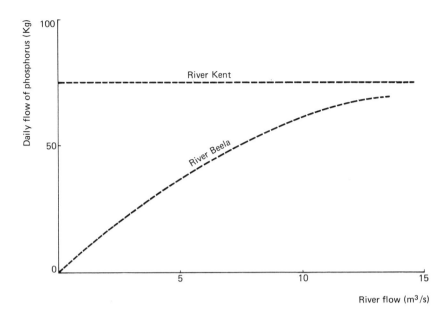

Figure 11.2 Comparison of phosphorus flow in two rivers which will contribute to the proposed Morecambe Bay reservoirs (simplified from Hoather, 1972)

ing on the seasonal pattern. The picture was quite different for the River Beela, where the daily flow of phosphorus was generally much lower and varied in relation to the total flow of water (Figure 11.2).[5] The consistent daily input of dissolved phosphorus into the River Kent has now been pin-pointed—its source is the effluent of the Kendal sewage works. If this effluent is accepted into the final scheme, which it probably must be, special treatment plant to precipitate and remove its phosphate content will be very desirable.

In order to regulate algal growth, and for environmental reasons (e.g. the preservation of estuarine life), a modified barrage scheme at Morecambe Bay is also a possibility. This would consist of two river-reservoirs enclosed by barriers and pump-feeding constructed circular reservoirs. In this system the circular reservoirs could accept a selected water mixture, the composition of which would probably vary on a seasonal basis.

Current feeling is that fish-stocking in reservoirs need not be deleterious to their use as water supplies, providing suitable controls are maintained. The Farmoor reservoir near Oxford has been stocked with both rainbow and brown trout and thus provides a recreational asset for anglers. If the modified barrage scheme at Morecambe Bay is implemented it is anticipated that the large eastern reservoir will be exploited equally.

References

1. J.N. COUCH, 1972. *Proc. Natn. Acad. Sci. USA* **69**, 2043-7
2. J.I.W. ANDERSON, 1974. *Salm. Trout Mag.* **200**, 28-37.
3. HMSO, 1972. *Morecambe Bay: Estuary Storage.*
4. J. CORLETT, 1972. *Wat. Treat. Exam.* **21**, 119-26.
5. R.C. HOATHER, 1972. *Wat. Treat. Exam.* **21**, 104-118.

Index

Alchromatium, 126
Achyranthes, 38
actinomycetes, 36, 113, 115
Actinophrys sol, 46, 47
Actinoplanes, 115
activated sludge process, 48, 138
aeration excursions, 49
aeration of rivers, 100, 101
aerenchyma tissue, 38
Aeromonas salmonicida, 90
Agapetus, 94-6; algal food of, 94
Alaska, Gulf of, 78
alevin, 62
algae: in benthic mud, 29; buoyancy, 31; communities, 29; ^{14}C assimilation by, 26; epilithic, 28; epipelic, 28; epiphytic, 28, 30; excretion, 20; extracellular products, 20; heterotrophic, 20; in Great Lakes, 27, 28; inherent growth characteristics, 28; microstraining, 134; nitrogen fixation, 32, 33; polyvalent viruses of, 36; spring maximum, 21; toxins, 35, 36; vacuole volume, 31; vitamin requirement, 25
algal food, 54, 94
ammonia: controlling *Prymnesium*, 36; in organic effluent, 144; in sewage, 138; toxicity to fish, 147
Amoeba, 22, 38, 82
Amphinemura sulcicollis (nymph), 96, 100
Anabaena circinalis, 31

Anabaena, 22, 38, 82
anaerobiosis, 24, 151, 154; conditions for, 124, 126, 141, 151
Aphanizomenon, 22, 34
Aqualinderalla fermentans, 116
Asellus, 95, 96, 102
Asterionella, 18, 21-5, 27; in lake sediments, 155; self-shading, 18; silicon requirement, 23, 24
Atlantic salmon, 75, 83, 88, 91, 159
autotrophic metabolism, 19

back-calculated growth, 71
bacteria: amylase-producers, 121; heterotrophic, 118–24; lipase-producers, 121; nitrogen fixation, 125; photosynthetic, 126; protease-producers, 121
Baetis (nymph), 82, 95, 96
Beggiatoa, 126
biochemical oxygen demand: of organic effluent, 144; of rivers, 101; of sewage, 48, 140
biological productivity, 16, 65
Blastocladia pringsheimii; 116
Blåsjön, Lake (Sweden), 86, 87
Bodo, 47, 48
body scales, 67, 78
Bosmina, 62, 76, 153
brackish water, 36

Caenis moesta (nymph), 96, 97
calcium: bicarbonate, 14; carbonate, 15
Callitriche, 38

Canadian high-arctic, 26
carbon dioxide, dissolved, 15
carbon fixation, 27
carbonic acid, 14
Carchesium, 46-8
carnivorous: animals, 46; ciliates, 48
carp, 84; grass, 83
cellulose decomposition, 111
Ceratium, 17, 22
Chaoborus: C.flavicans, 58-60; larva, 51
char, 63, 85; attractant, 80; homing, 80; Norwegian, 80
Chara, 30, 38
chironomid midge larvae, 76, 81, 102, 144
Chironomus, 153; *C. thummi*, 102
Chironomus anthracinus, 53-8; algal food, 54; larval stages, 54-7; predation by fish, 56, 57
chitin decomposition, 113-16; 119
Chlorella, 20, 22
chloride: river content, 103; chlorination, 136
chlorophylls, 17, 127, 152, 160; chlorophyll-a values, 27, 160
cholera, 133
Chromatium, 122, 125, 127
Chydorus, 62
Chytridiales, 114
ciliates, 44, 46-9, 143
Cladophora, 44, 45, 94, 194
Clear Lake (California), 33
Coleochaete, 30
colonization: by vascular plants, 40; by invertebrates, 95
copper (toxicity to fish), 147
Corixa.dorsalis, 96, 97
Cultus Lake, 77
cyanophage, 36
Cyclops strenuus abyssorum, 51, 59
Cyclops, 50, 62, 76, 86, 153

Daphnia, 50, 76, 153
Dalnee, Lake (Kamchatka), 76
dating, *see* lake sediments

DDT, 146, 158
Dee estuary, 159
denitrification, 124; by micro-flora, 125
deoxygenation: of lakes, 13; of R.Rhine, 102; of R.Thames, 141
detergents: anionic, 139; soft, 140
Diaptomus, 50, 86; *D.arcticus*, 52
dinoflagellates, 17
'drift', 81

edaphic climax, 41
effluent: farm, 16; farmyard, 48; organic, 144, 145
Eichhernia crassipes, 42
Eleocharis, 38
Elodea, 38, 62
epilimnion, 10
escapement ratio, 78
Escherichia coli, 133
Esox lucius, 61-74
Esrom, Lake (Denmark), 11, 12, 53

Farmoor reservoir, 161
fermentative fungi, 116—18
fish: antibodies in, 91; disease, 89-92 -ponds, 36, 84; *Saprolegnia* disease, 91; sulphonamide drugs, 91; viral haemorrhagic septicaemia, 89
fish serum, 88; electrophoresis, 88; transferrins in, 89
fluorescent antibody stain, 90
Fontinalis, 18, 38-40, 98
Ford-Walford plot, 70
Fraser River (Canada), 77
freshets, 80
furunculosis, 90

Gammarus, 63, 81, 85, 97
Gammarus pulex, 97, 98; calcium for, 97
George, Lake (Uganda), 25, 51
glycollic acid, 20
grazing, 24, 31
Great Lakes (North America), 27, 28

herbicides, 42, 146
heterocysts, 22, 32, 33
heterotrophs: algae, 20: bacteria, 118-24
homing: of char, 80; of salmon, 79
humic substances, 17, 118
hydrogen sulphide, 30, 141
hypolimnion, 10, 23, 49, 120, 126, 129

ice, 9
ichthyocides, 85
Ilybius, 96, 97
growth: characteristics, 28; optimum growth temperature, 9
insulation (ponds and lakes), 9
iron, 150; -cycle, 128
Ishnura elegans (nymph), 96, 97

Japanese Lakes, 18

Kariba Dam, 42
kokanee, 86
Kootenay Lake (Canada), 86

lakes: Blåsjön, 86, 87; Dalnee, 76; Esrom, 11, 12, 53; George, 25, 51; insulation in, 9; Kootenay, 86; mixing in, 10; oligotrophic, 34; overturn, 11; pH rise, 18; productivity, 16, 150; Vechten, 120
lake sediments; ^{14}C dating, 149, 154; caesium-137 dating, 154; iron in, 150; magnetic dating, 154; manganese in, 150, 151
larval stages, 54-7, 145, 146, 158
lead (toxicity to fish), 147
leaf decomposition, 108-13
Lemna, 42
Leven, Loch (Scotland), 51
Limnaea pereger, 95, 96
Littorella, 38-41
Lobelia, 38
Loxodes, 46-9; 'aeration excursions', 49

Mackereth sediment sampler, 154, 155

magnetic dating, 154
Malham Tarn, 30
manganese, 150, 151
mayfly nymphs, 81, 95, 144; 'drift', 81
Melosira, 21, 22, 25, 27, 28
methane oxidizers, 129
Microcystis, 22, 25, 35, 38, 82
microstraining (algae), 134
'migrating disposition', 76
mixing: of lakes, 10; of reservoirs, 133
Morecambe Bay, 159
mosquito larvae, 145, 146, 158
Myriophyllum, 62; *M. spicatum,* 39, 93

Ness, Loch, 41
net selection, 65
Nitella, 18, 38, 62
nitrate hazard, 134
nitrogen:-cycle, 124-6; fixation, 32, 125; in fishponds, 84, 85; transfer, 112, 113
Nitrobacter, 124
Nitrosomonas, 124
Nuphar, 38-40

Onchorhynchus nerka, 61, 75-80, 86
opercular bone, 68
oxygen (supersaturation in water), 12, 19

Pacific salmon, 75; body scales, 78; spawning, 75
paddy fields, 32
Paramecium, 46-8
pesticides, 146
pH: for *Gammarus pulex,* 97; of water, 15; rise in lakes, 18
phosphatase enzymes, 130, 131
phosphorus: circulation, 129; extraction from sewage, 138, 161; flow in rivers, 161
photic zone, 17, 21, 27
Phragmites reeds, 30, 40, 41
Pisidium, 57, 96
Pythium, 109

Phytophthora, 109
pike: alevin, 62; back-calculated growth, 71; body scales, 67; fecundity, 64; net selection, 65; opercular bone, 68; recapture, 69, 71; spawning, 61
pike-perch, 83
plankton, 77
polar night, 26
poliomyelitis, 133
pollution: organic, 31
Potamogeton, 38-41
Prymnesium parvum, 22, 35; control, 36

Ranunculus fluitans, 39, 93
respiration, 19; in community, 58
Rhine, River, 102; chloride content of, 103

salmon, 74, 83; Atlantic, 75; escapement ratio, 78; in fish farming, 159; fingerlings, 76; homing, 79; 'migrating disposition' 76; Pacific, (*see* heading); pink, 80; redd 75; serological differentiation, 78; smolts, 76, 159; sockeye, 78; spawning, 75; thyroid activity, 76
Salmo trutta, 63
Salvelinus: S. alpinus, 63; *S. fontinalis,* 90
Salvina auriculata, 42
San Antonio, Texas, USA, 138
saprobiological triangle, 142
Saprolegnia, 104; disease of fish, 91
Scenedesmus, 22, 30, 31; as ciliate food, 47-8
serological differentiation, 78
sewage, 26, 48; activated sludge process, 48, 138; effluent, 16, 101, 140; fungus, 143, 144; human, 16; percolating filter, 139; phosphorus extraction, 138, 161; sedimentation, 137, 140; sludge as fertilizer, 138
silicon requirement, 23, 24
soil (under water), 40
spawning: of *Esox lucius,* 61; of *Onchorhynchus nerka,* 75
Sphaerotilus natans, 122, 128, 143, 145; nutrient extraction by, 145
spring maximum, 21
Stigeoclonium, 94; in organic effluent, 144
stonefly nymphs, 81, 95, 144
stream-bed colonization, 95
succession, protozoan, 48
sudd, 43
sulphonamide drugs, 91
sulphur cycle, 126

Tabellaria, 22, 30
Thames, River, 140, 141
thermocline, 10
thyroid activity, 76
Tilapia nilotica, 82
toxicity, copper, 147; lead, 147; zinc, 147
toxins (algal): exo-, 35, 36; endo-, 35; co-factor, 36
trout, 63; brook, 90; Egtved disease, 89; rainbow, 83, 89, 90
Tubifex tubifex, 96, 97
tubificid worms, 57, 97, 102, 144
typhoid fever, 133

ultra-violet: inactivation effect, 36; sterilizing effect, 18

Vechten, Lake (Netherlands), 120
Victoria Falls, 42
viruses, of algal blooms, 25; as human pathogens in water, 133, 135; plaques, 37; polyvalent, 36; septicaemia of fish, 89
Von Bertalanffy equation, 69
Vorticella, 46-8; as biological indicator, 48, 142

Wash, the, 159
water: density, 9; domestic use of, 132; coliform count, 134; hard, 14; hardness of, 15, 147; pH, 15; sand-filtration of, 133, 134; supersaturation of oxygen in, 12, 19

Windermere, Lake, 21, 28, 154
wood decomposition, 111

Zambesi River, 42
zinc (toxicity to fish), 147